I0488210

Fluvial Transport of Mercury, Dissolved Organic Carbon, Suspended Sediment, and Selected Major Ions in Contrasting Stream Basins in South Carolina and New York, October 2004 to September 2009

By Celeste A. Journey, Douglas A. Burns, Karen Riva-Murray, Mark E. Brigham, Daniel T. Button, Toby D. Feaster, Matthew D. Petkewich, and Paul M. Bradley

National Water-Quality Assessment Program

Scientific Investigations Report 2012–5173

U.S. Department of the Interior
U.S. Geological Survey

U.S. Department of the Interior
KEN SALAZAR, Secretary

U.S. Geological Survey
Marcia K. McNutt, Director

U.S. Geological Survey, Reston, Virginia: 2012

For more information on the USGS—the Federal source for science about the Earth, its natural and living resources, natural hazards, and the environment, visit http://www.usgs.gov or call 1–888–ASK–USGS.

For an overview of USGS information products, including maps, imagery, and publications, visit http://www.usgs.gov/pubprod

To order this and other USGS information products, visit http://store.usgs.gov

Suggested citation:
Journey, C.A., Burns, D.A., Riva-Murray, Karen, Brigham, M.E., Button, D.T., Feaster, T.D., Petkewich, M.D., and Bradley, P.M., 2012, Fluvial transport of mercury, organic carbon, suspended sediment, and selected major ions in contrasting stream basins in South Carolina and New York, October 2004 to September 2009: U.S. Geological Survey Scientific Investigations Report 2012–5173, 125 p.

Foreword

The U.S. Geological Survey (USGS) is committed to providing the Nation with reliable scientific information that helps to enhance and protect the overall quality of life and that facilitates effective management of water, biological, energy, and mineral resources (*http://www.usgs.gov/*). Information on the Nation's water resources is critical to ensuring long-term availability of water that is safe for drinking and recreation and is suitable for industry, irrigation, and fish and wildlife. Population growth and increasing demands for water make the availability of that water, measured in terms of quantity and quality, even more essential to the long-term sustainability of our communities and ecosystems.

The USGS implemented the National Water-Quality Assessment (NAWQA) Program in 1991 to support national, regional, State, and local information needs and decisions related to water-quality management and policy (*http://water.usgs.gov/nawqa*). The NAWQA Program is designed to answer: What is the quality of our Nation's streams and groundwater? How are conditions changing over time? How do natural features and human activities affect the quality of streams and groundwater, and where are those effects most pronounced? By combining information on water chemistry, physical characteristics, stream habitat, and aquatic life, the NAWQA Program aims to provide science-based insights for current and emerging water issues and priorities. From 1991 to 2001, the NAWQA Program completed interdisciplinary assessments and established a baseline understanding of water-quality conditions in 51 of the Nation's river basins and aquifers, referred to as Study Units (*http://water.usgs.gov/nawqa/studies/study_units.html*).

National and regional assessments are ongoing in the second decade (2001–2012) of the NAWQA Program as 42 of the 51 Study Units are selectively reassessed. These assessments extend the findings in the Study Units by determining water-quality status and trends at sites that have been consistently monitored for more than a decade, and filling critical gaps in characterizing the quality of surface water and groundwater. For example, increased emphasis has been placed on assessing the quality of source water and finished water associated with many of the Nation's largest community water systems. During the second decade, NAWQA is addressing five national priority topics that build an understanding of how natural features and human activities affect water quality, and establish links between *sources* of contaminants, the *transport* of those contaminants through the hydrologic system, and the potential *effects* of contaminants on humans and aquatic ecosystems. Included are studies on the fate of agricultural chemicals, effects of urbanization on stream ecosystems, bioaccumulation of mercury in stream ecosystems, effects of nutrient enrichment on aquatic ecosystems, and transport of contaminants to public-supply wells. In addition, national syntheses of information on pesticides, volatile organic compounds (VOCs), nutrients, trace elements, and aquatic ecology are continuing.

The USGS aims to disseminate credible, timely, and relevant science information to address practical and effective water-resource management and strategies that protect and restore water quality. We hope this NAWQA publication will provide you with insights and information to meet your needs, and will foster increased citizen awareness and involvement in the protection and restoration of our Nation's waters.

The USGS recognizes that a national assessment by a single program cannot address all water-resource issues of interest. External coordination at all levels is critical for cost-effective management, regulation, and conservation of our Nation's water resources. The NAWQA Program, therefore, depends on advice and information from other agencies—Federal, State, regional, interstate, Tribal, and local—as well as nongovernmental organizations, industry, academia, and other stakeholder groups. Your assistance and suggestions are greatly appreciated.

William H. Werkheiser
USGS Associate Director for Water

Contents

Figures

Figures—Continued

Tables

Conversion Factors

Inch/Pound to SI

Multiply	By	To obtain
Length		
inch (in.)	2.54	centimeter (cm)
inch (in.)	25.4	millimeter (mm)
foot (ft)	0.3048	meter (m)
mile (mi)	1.609	kilometer (km)
mile, nautical (nmi)	1.852	kilometer (km)
yard (yd)	0.9144	meter (m)
Area		
acre	4,047	square meter (m^2)
acre	0.4047	hectare (ha)
acre	0.4047	square hectometer (hm^2)
acre	0.004047	square kilometer (km^2)
square foot (ft^2)	929.0	square centimeter (cm^2)
square foot (ft^2)	0.09290	square meter (m^2)
square inch (in^2)	6.452	square centimeter (cm^2)
section (640 acres or 1 square mile)	259.0	square hectometer (hm^2)
square mile (mi^2)	259.0	hectare (ha)
square mile (mi^2)	2.590	square kilometer (km^2)
Volume		
cubic inch (in^3)	16.39	cubic centimeter (cm^3)
cubic inch (in^3)	0.01639	cubic decimeter (dm^3)
cubic inch (in^3)	0.01639	liter (L)
cubic foot (ft^3)	28.32	cubic decimeter (dm^3)
cubic foot (ft^3)	0.02832	cubic meter (m^3)
cubic yard (yd^3)	0.7646	cubic meter (m^3)
cubic mile (mi^3)	4.168	cubic kilometer (km^3)
acre-foot (acre-ft)	1,233	cubic meter (m^3)
acre-foot (acre-ft)	0.001233	cubic hectometer (hm^3)
Flow rate		
acre-foot per day (acre-ft/d)	0.01427	cubic meter per second (m^3/s)
acre-foot per year (acre-ft/yr)	1,233	cubic meter per year (m^3/yr)
cubic foot per second (ft^3/s)	0.02832	cubic meter per second (m^3/s)
cubic foot per second per square mile [(ft^3/s)/mi^2]	0.01093	cubic meter per second per square kilometer [(m^3/s)/km^2]
cubic foot per day (ft^3/d)	0.02832	cubic meter per day (m^3/d)
gallon per minute (gal/min)	0.06309	liter per second (L/s)
gallon per day (gal/d)	0.003785	cubic meter per day (m^3/d)
gallon per day per square mile [(gal/d)/mi^2]	0.001461	cubic meter per day per square kilometer [(m^3/d)/km^2]
million gallons per day per square mile [(Mgal/d)/mi^2]	1,461	cubic meter per day per square kilometer [(m^3/d)/km^2]

Conversion Factors—Continued

SI to Inch/Pound

Multiply	By	To obtain
Length		
centimeter (cm)	0.3937	inch (in.)
kilometer (km)	0.6214	mile (mi)
kilometer (km)	0.5400	mile, nautical (nmi)
Area		
square kilometer (km^2)	247.1	acre
hectare (ha)	0.003861	square mile (mi^2)
square kilometer (km^2)	0.3861	square mile (mi^2)
Volume		
liter (L)	33.82	ounce, fluid (fl. oz)
liter (L)	2.113	pint (pt)
liter (L)	1.057	quart (qt)
liter (L)	0.2642	gallon (gal)
cubic meter (m^3)	264.2	gallon (gal)
cubic meter (m^3)	0.0002642	million gallons (Mgal)
cubic meter (m^3)	35.31	cubic foot (ft^3)
cubic meter (m^3)	1.308	cubic yard (yd^3)
cubic meter (m^3)	0.0008107	acre-foot (acre-ft)
Flow rate		
cubic meter per second (m^3/s)	70.07	acre-foot per day (acre-ft/d)
cubic meter per second (m^3/s)	35.31	cubic foot per second (ft^3/s)
cubic meter per second per square kilometer [(m^3/s)/km^2]	91.49	cubic foot per second per square mile [(ft^3/s)/mi^2]
cubic meter per day per square kilometer [(m^3/d)/km^2]	684.28	gallon per day per square mile [(gal/d)/mi^2]
Mass		
gram (g)	0.03527	ounce, avoirdupois (oz)
kilogram (kg)	2.205	pound, avoirdupois (lb)
Application rate		
kilograms per hectare per year [(kg/ha)/yr]	0.8921	pounds per acre per year [(lb/acre)/yr]

Temperature in degrees Celsius (°C) may be converted to degrees Fahrenheit (°F) as follows:
°F=(1.8 ×°C)+32

Temperature in degrees Fahrenheit (°F) may be converted to degrees Celsius (°C) as follows:
°C=(°F 32)/1.8

Horizontal coordinate information is referenced to the North American Datum of 1983 (NAD 83).

Concentrations of chemical constituents in water are given in milligrams per liter (mg/L), micrograms per liter (µg/L), or nanograms per liter (ng/L).

x

Fluvial Transport of Mercury, Dissolved Organic Carbon, Suspended Sediment, and Selected Major Ions in Contrasting Stream Basins in South Carolina and New York, October 2004 to September 2009

By Celeste A. Journey, Douglas A. Burns, Karen Riva-Murray, Mark E. Brigham, Daniel T. Button, Toby D. Feaster, Matthew D. Petkewich, and Paul M. Bradley

Abstract

A spatially extensive assessment of the environmental controls on mercury transport and bioaccumulation in stream ecosystems in New York and South Carolina was conducted as part of the U.S. Geological Survey National Water-Quality Assessment Program and included the determination of fluvial transport of mercury and associated constituents during water years 2005–2009. (A water year extends from October of one calendar year to September of the next calendar year.) In the Coastal Plain region of South Carolina, the study area included the Edisto River and its headwater tributary, McTier Creek. In the Adirondack region of New York, the study area included the upper Hudson River and its headwater tributary, Fishing Brook. Median concentrations of filtered total mercury ranged from 1.55 nanograms per liter (ng/L) at the Hudson River site to 2.77 ng/L at the Edisto River site. The Edisto River site had the greatest median filtered methylmercury concentration, at 0.32 ng/L, and the Hudson River site had the least median filtered methylmercury concentration, at 0.07 ng/L.

Two-year (2008 and 2009) mean annual filtered methylmercury yield at the McTier Creek site of 0.025 microgram per square meter per year [(μg/m^2)/yr] was almost 3 times less than the filtered methylmercury yield of 0.064 (μg/m^2)/yr at the Edisto River site, indicative of increased contributions with increasing scale from headwater stream catchment to larger river basin. Two-year mean yields of filtered total mercury were 0.83 and 0.68 (μg/m^2)/yr at the McTier and Edisto sites, respectively, indicative of negligible change in contribution from headwater stream to larger river scale. Yields of particulate forms of mercury were relatively consistent between headwater stream and larger river basin scales. Two-year mean annual yields of particulate methylmercury were 0.009 and 0.011 (μg/m^2)/yr at the McTier Creek and Edisto River sites, respectively. Annual particulate total mercury yields were 0.34 (μg/m^2)/yr at the McTier Creek site and 0.27 (μg/m^2)/yr at the Edisto River site. In contrast to the South Carolina sites, the 2-year mean filtered methylmercury yield of 0.095 (μg/m^2)/yr at the Fishing Brook site in New York was higher than the filtered methylmercury yield of 0.068 (μg/m^2)/yr at the Hudson River site, indicating decreased contributions with increasing scale from headwater stream catchment to larger river basin. As observed with the South Carolina sites, the 2-year mean yields of filtered total mercury indicated no change in contribution from headwater stream to larger river scale and were 1.67 and 1.66 (μg/m^2)/yr at the Fishing Brook and Hudson sites, respectively.

Mean annual dissolved organic carbon yields of 24.4 and 16.7 kilograms per hectare per year at the McTier Creek and Edisto River paired basin sites, respectively, were lower than the mean annual dissolved organic carbon yields of 54.4 and 52.9 kilograms per hectare per year at the Fishing Brook and Hudson River paired basin sites, respectively. In South Carolina, mean annual dissolved chloride yields increased slightly with basin scale. Conversely, in New York, mean annual dissolved chloride yields decreased slightly with basin scale; however, basin-scale changes in mean annual dissolved sulfate yields were more consistent between the two paired basins and indicated increasing yields with increasing basin scales. Mean annual suspended sediment yields did not exhibit a consistent pattern between the New York and South Carolina basins or between paired (headwater-large river) sites.

In the McTier Creek headwater stream basin, mean annual wet deposition of total mercury was 9.91 (μg/m^2)/yr for water years 2005 through 2009. In 2007, litterfall accounted for an estimated 12.8 (μg/m^2)/yr of total mercury in this basin. Based on the 2007 estimates, wet deposition of total mercury represented only 37 percent of the total mercury deposition in the McTier Creek basin, and only 7 percent of the total mercury deposition in the McTier Creek basin reached the stream site, indicating storage of the atmospherically derived total mercury within the basin. Mean annual wet deposition of total mercury was 6.30 (μg/m^2)/yr for water years 2005 through 2009 in the Fishing Brook basin. Litterfall accounted for an estimated 9.00 (μg/m^2)/yr of total mercury in this basin in 2007, which produced a total mercury deposition of 15.4 (μg/m^2)/yr. Based on the 2007 estimates, wet deposition of total mercury represented only 42 percent of the total mercury deposition in the Fishing Brook basin. In 2007, only 13 percent of the total mercury deposition in the Fishing Brook basin reached the stream site, indicating storage of the atmospherically derived total mercury within the basin.

Introduction

Mercury contamination in fish is a major concern globally and is the leading national cause of fish consumption advisories issued by States for the protection of human health. In fact, the U.S. Environmental Protection Agency (USEPA) summarized that 3,361 fish consumption advisories in 2008 were issued by States for mercury, affecting 16.8 million lake acres and 1.3 million river miles (U.S. Environmental Protection Agency, 2009). The 2008 mercury advisories demonstrated an increase from 3,080 advisories in 2006. Mercury contamination was responsible for almost 80 percent of all fish consumption advisories issued by States in 2008 (U.S. Environmental Protection Agency, 2009). Wet and dry atmospheric deposition are often the major sources of mercury in aquatic and terrestrial ecosystems (U.S. Environmental Protection Agency, 1997; National Research Council, 2000).

Certain physical and geochemical conditions within stream basins have been reported to be conducive for net increases or decreases of total mercury (THg) and the more biologically available methylmercury (MeHg) concentrations in streamwater and fish tissue. The terrestrial environment is an important component of mercury cycle and bioaccumulation processes in streams and rivers because of its role in the transformation of atmospherically derived mercury (Hg) to the more bioavailable MeHg form and delivery of THg (includes all inorganic and organic forms) and MeHg to streams (Hurley and others, 1995; Kolka and others, 1999; Grigal, 2002; Gabriel and Williamson, 2004; Brigham and others, 2009; Bradley and others, 2011). Thus, bioaccumulation of Hg in fish tissues depends not only on microbially mediated production and persistence of MeHg but also on the transport of MeHg from the site of environmental production to the point of entry into the food web and on the efficiency of biomagnification of MeHg within the food web (Watras and others, 1994; Downs and others, 1998; Hammerschmidt and Fitzgerald, 2006; Munthe and others, 2007; Bradley and others, 2009). Transformation of inorganic forms of Hg (Hg^{2+}, Hg^o) to MeHg often occurs in the terrestrial environment, especially within wetlands and riparian floodplains (Grigal, 2002; Munthe and others, 2007). For most streams, Hg and MeHg are delivered to streams by two possible mechanisms: (1) atmospheric deposition directly to the stream (mainly inorganic forms of Hg) or (2) indirectly, by hydrologic transport of atmospherically derived Hg from the terrestrial environment to the stream (Grigal, 2002, 2003; Hammerschmidt and Fitzgerald, 2006; Munthe and others, 2007; Risch and others, 2012). The dissolved and particulate-bound forms of Hg and MeHg can be delivered to streams by hydrologic transport and should be considered when quantifying fluvial transport of Hg.

Both wet and dry deposition of atmospherically derived Hg play a role in Hg delivery to aquatic or terrestrial environments (Risch and others, 2012). Wet deposition of Hg includes Hg delivered to land and water surfaces during precipitation events, including rain, sleet, hail, and fog. Dry deposition of Hg is the continuous transfer of Hg from the atmosphere to plants, soil, water, and snow. Because forest canopies can accumulate Hg, the mass of Hg in litterfall is reported to represent a large portion of the overall atmospheric deposition of Hg to the environment (Johnson and Lindberg, 1995; Grigal, 2002; Risch and others, 2012). In the eastern United States, annual dry deposition of Hg from litterfall was substantially higher than annual wet deposition of Hg (Risch and others, 2012).

Because of the strong affinity of Hg for organic matter (Haitzer and others, 2002), the fate and transport of Hg is largely controlled by its interactions with dissolved organic carbon (DOC) and suspended particles, especially particulate organic matter (Yin and Balogh, 2002; Brigham and others, 2009). Many studies have reported on the strong relation between filtered total mercury (FTHg; dissolved and colloidal phases of the inorganic and methylated species) and concentrations and character of DOC (Babiarz and others, 2001; Haitzer and others, 2002; Skyllberg and others, 2003; Driscoll and others, 2007; Brigham and others, 2009; Dittman and others, 2009; Glover and others, 2010). Mercury and DOC transport in surface waters are controlled ultimately by water flux and hydrologic flowpaths. Areas that are characterized by high DOC accumulation, such as riparian floodplains and wetlands, are likely zones where Hg-DOC complexes are mobilized to streamwater, especially during high-flow events (Grigal, 2002). Therefore, quantifying DOC flux can contribute to the understanding of the fate and transport of Hg in surface-water systems. Estimates of Hg and DOC loads (flux) in streams are needed to assess the fate of Hg deposited within a watershed and to provide a measure of watershed response to changes in Hg emissions (Lindqvist, 1991; Hurley, 1998; Kolka and others, 1999). Load estimation at multiple spatial scales within a river basin would provide a means to quantify and compare the ability of mercury and DOC to move through a surface-water system.

Previous Investigations

In 1998 and 2005, the U.S. Geological Survey (USGS) Toxic Substances Hydrology Program (Toxics) and the USGS National Water-Quality Assessment Program (NAWQA) conducted two reconnaissance-type studies and reported elevated Hg concentrations [mean of 1.0 plus or minus ($\pm$) standard deviation of 0.46 microgram per gram ($\mu g/g$) wet weight] in largemouth bass in the Edisto River basin in South Carolina (S.C.; Brumbaugh and others, 2001; Bauch and others, 2009; Scudder and others, 2009). The mean Hg concentration in largemouth bass in the Edisto River reported by the NAWQA studies was much higher than Hg concentrations in largemouth bass collected by the South Carolina Department of Health and Environmental Control (SCDHEC) in streams in the Congaree National Park located in the adjacent Congaree

River basin (mean of 0.2 ± standard deviation of 0.1 μg/g) (Bradley and others, 2009; Bradley and others, 2010). Similar riparian wetlands coverage and a similar range of sediment Hg methylation rates in the Edisto and Congaree study areas could not explain the difference in fish tissue Hg concentrations (Bradley and others, 2009). A second study of the hydrology (groundwater/surface-water interaction) of these two study areas evaluated the connection of the river systems to the MeHg sources (wetlands) and the ability of MeHg to be transported from its terrestrial (wetland) source area to the river (Bradley and others, 2010). The study indicated that the substantial differences in Hg bioaccumulation between the Congaree and Edisto River systems could be explained by better efficiency of MeHg transport from terrestrial source areas to the river in the Edisto River basin compared to the Congaree rather than to MeHg production and persistence (Bradley and others, 2010). The role of MeHg transport from the terrestrial to the fluvial environment is further supported by the work of Glover and others (2010), who reported a link between river regulation by dams (indicative of loss of connection with the floodplain) and lower modeled Hg concentration in largemouth bass (*Micropterus salmoides*) in South Carolina rivers. In a basin-wide reconnaissance of unfiltered MeHg concentrations in the Edisto River, MeHg concentrations increased with distance downstream consequent with increased wetland densities (from 5 to 10 percent in headwater basins such as McTier Creek to greater than 20 percent farther downstream in the Edisto River basin) (Bradley and others, 2011).

The Adirondack region of New York is considered a mercury "hot spot," indicating widespread high levels of Hg bioaccumulation relative to surrounding regional landscapes, based on the presence of controlling factors such as the high amounts of forest and wetland cover and an abundance of high dissolved organic matter and low-pH streams and lakes (Driscoll and others, 2007; Evers and others, 2007; Simonin and others, 2008). During previous studies within the upper Hudson River basin, high levels of THg and MeHg have been detected in streams and lakes, and the biogeochemical processes and landscape interactions that affect Hg concentrations and loads in these waters were explored (Bushey and others, 2008a; Dittman and others, 2009, 2010; Selvendiran and others, 2008a, 2008b, 2009; Schelker and others, 2011). This work in the Upper Hudson has also quantified key fluxes and pools of Hg in Adirondack watersheds (Bushey and others, 2008b; Choi and others, 2008).

From 2002 to 2006, the USGS NAWQA Program conducted a spatially extensive assessment of the environmental controls on Hg transport and bioaccumulation in stream ecosystems in the United States. In addition to Hg, concentrations of DOC, suspended sediment (SSC), nutrients, and major ions were monitored in streamwater, streambed sediment, pore water, and selected biota in stream basins in Oregon, Wisconsin, and Florida (Brigham and others, 2009; Chasar and others, 2009; Marvin-DiPasquale and others, 2009). Selected stream

basins had large ranges in environmental conditions, including climate, landscape characteristics, atmospheric deposition of Hg (predominant source in all basins), and water chemistry. A major finding of these investigations was that environmental conditions within the selected stream basins provided controls on net methylation of Hg and fluvial transport of total and methylated species of Hg. Specifically, wetland density, DOC and SSC concentrations, and streamflow were reported to be correlated to Hg species in water (Brigham and others, 2009).

In 2005, further assessment on the potential of small-scale [drainage areas of less than 50 square miles (mi²)] headwater streams to transform and transport mercury to large-scale river basins was recognized as a need. Synoptic surveys of THg and MeHg were conducted in the Edisto and Hudson River basins to provide information for the selection of intensive monitoring locations (core sites), and results were described in Bradley and others (2011). Because significant MeHg concentrations and limited spatial variability were reported throughout the Edisto River basin, McTier Creek near New Holland, S.C. (McTier Creek, USGS station number 02172305; fig. 1; table 1) was selected as the core site (location of intensive sampling and monitoring) in the Edisto River basin. The selected core site is a tributary to the South Fork of the Edisto River and is located just downstream of a long-term streamflow, biological, and water-quality monitoring location (McTier Creek near Monetta, S.C.; USGS station number 02172300; table 1) (Bradley and others, 2011; Scudder Eikenberry and others, 2011). Additionally, Edisto River near Givhans, S.C. (Edisto River; station 02175000) was included in the mercury assessment for comparison purposes as a larger scale river basin, but water-quality and biological monitoring were more focused on the smaller McTier Creek basin (fig. 1; table 1). Fishing Brook (County Line Flow) near Newcomb, N.Y. (Fishing Brook; USGS station number 0131199050) was selected for the core site in the upper Hudson River basin (Bradley and others, 2011; Scudder Eikenberry and others, 2011) (fig. 1; table 1). Fishing Brook is a tributary to the upper Hudson River and is located upstream of the Hudson River near Newcomb, N.Y. (Hudson River; USGS station number 01312000), which also was included as a larger scale, but more limited, monitoring site. Information on hydrology (Bradley and others, 2009; Feaster and others, 2010; Benedict and others, 2011; Schelker and others, 2011), spatial and seasonal variability of MeHg in streamwater (Bradley and others, 2009; Bradley and others, 2011; Schelker and others, 2011), spatial variability of Hg in macroinvertebrates and fish (Riva-Murray and others, 2011), spatial variability of Hg in soils (Woodruff and others, 2011), and environmental characteristics (Scudder Eikenberry and others, 2011) has been published for the two core sites, McTier Creek and Fishing Brook. Previous publications have presented limited information about comparisons between Hg transport in the headwater stream core sites and Hg transport in larger river basin sites.

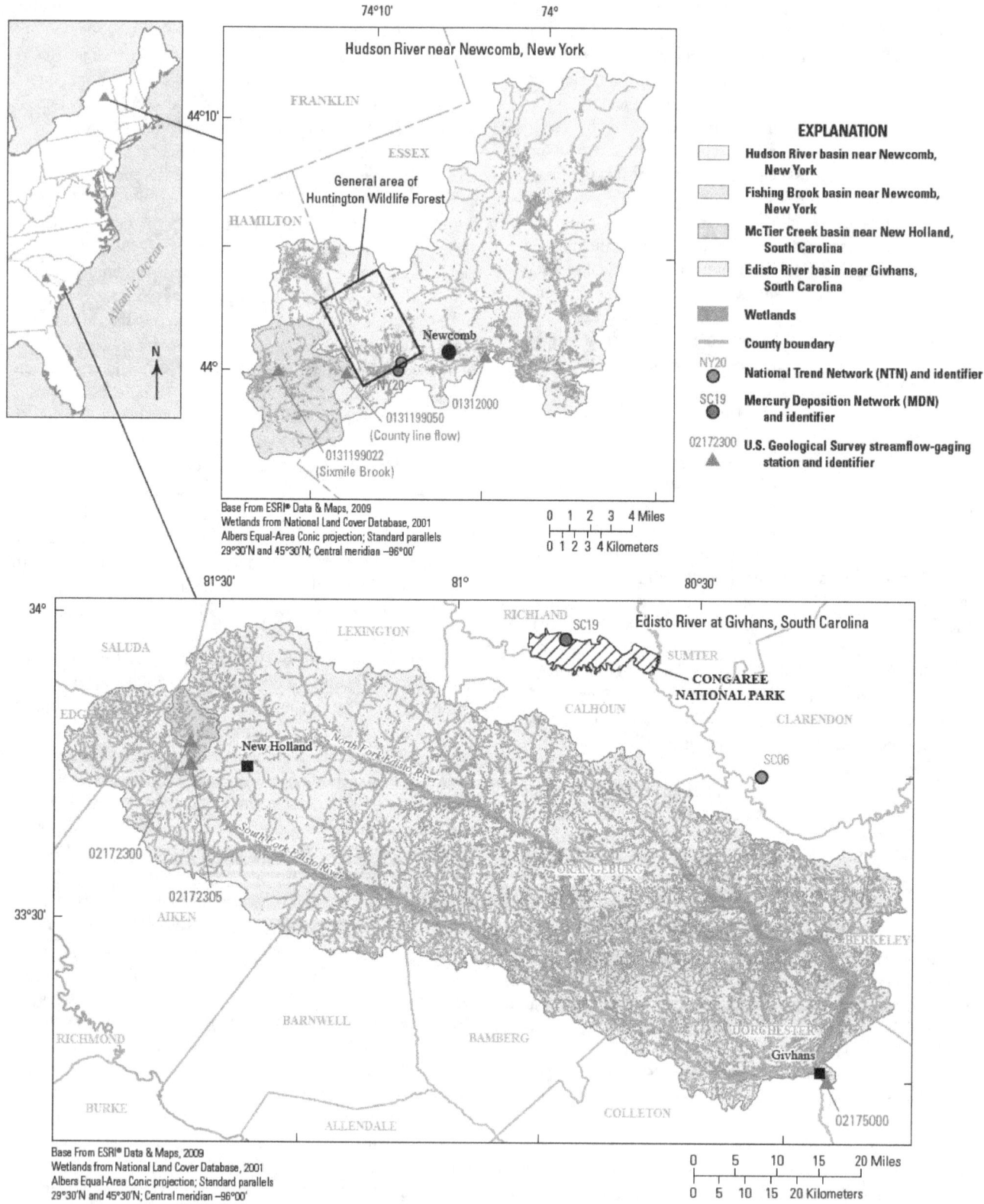

Figure 1. Overview map of the New York and South Carolina study areas: (top) the upper Hudson River basin with the headwaters subbasin of Fishing Brook, New York; (bottom) the Edisto River basin with the headwaters subbasin of McTier Creek, South Carolina.

Purpose and Scope

As part of the Mercury Topical Study of the USGS NAWQA Program, a spatially extensive assessment of the environmental controls on Hg transport and bioaccumulation in stream ecosystems was conducted in New York and South Carolina from 2005 to 2009. The purpose of this report is to describe the concentrations, loads, and yields of particulate MeHg (PMeHg), filtered MeHg (FMeHg), particulate THg (PTHg), FTHg, SSC, DOC, particulate organic carbon (POC), and selected major ions in the Edisto River basin at the McTier Creek and Edisto River sites and in the Upper Hudson River basin at the Fishing Brook and Hudson River sites (fig. 1; table 1). The objectives of the assessment that are discussed in this report are to (1) compare Hg and selected water-quality constituent concentrations among sites; (2) compute and compare loads and yields of Hg and selected water-quality constituents at headwater stream sites and large-scale river sites; and (3) evaluate the potential of small-scale [drainage areas of less than 120 square kilometers (km^2) (50 mi^2)] headwater streams to transform and transport mercury.

Three appendixes in this report contain the streamflow (appendix 1), water-quality (appendix 2), and load (appendix 3) data associated with the McTier Creek, Edisto River, Fishing Brook, and Hudson River sites and used in the data analysis. Additionally, appendixes 2-E and 2-F contain miscellaneous water-quality data that were collected within the watersheds of the four sites, but not directly used in the data analysis.

Study Area Description

As described previously, the study design paired two headwater streams with their larger river basins. In the Coastal Plain region of South Carolina, the study area included McTier Creek, which is a headwater tributary to the Edisto River (fig. 1; table 1). The McTier Creek basin is 79.5 km^2 and represents only 1.1 percent of the much larger Edisto River basin (7,071 km^2; table 1). In the Adirondack region of New York, the study area included Fishing Brook, a tributary to the upper Hudson River (fig. 1; table 1). The Fishing Brook basin is 65.3 km^2 and represents about 13 percent of the Hudson River basin at the study location (497 km^2; table 1). Environmental characteristics of the study basins are described in detail in Scudder Eikenberry and others (2011), so only a brief summary is provided in this report.

Edisto River Basin

The climate of this area in South Carolina is sub-tropical with relatively mild winters and distinct wet and dry seasons. The wet and dry seasons result in seasonally fluctuating water levels in riparian pools, wetlands, and streams of the area (Bradley and others, 2011). During the study period for the McTier Creek site, the mean winter air temperature was

9.07 degrees Celsius (°C), and the mean summer air temperature was 26.8 °C (Scudder Eikenberry and others, 2011). Mean annual precipitation averaged 113.4 centimeters per year (cm/yr) during the study period (Scudder Eikenberry and others, 2011).

The Edisto River basin drains 7,071 km^2 within the Southeastern Plains (SEP) (upper portion) and Middle Atlantic Coastal Plain (MACP) (lower portion) Level III Ecoregions in South Carolina as an unregulated (free flowing) river (Griffith and others, 2002; U.S. Environmental Protection Agency, 2002; Omernik, 2005). The Edisto River has low stream gradients, extensive riparian wetlands, and highly tannic (DOC) water, which is characteristic of streams in the SEP and MACP (Bradley and others, 2009). The Edisto River is formed by the confluence of the North and South Fork Edisto Rivers (fig. 1). McTier Creek is a headwater tributary of the South Fork Edisto River. For the Edisto River station (02175000), land use is mostly rural, with 35.9 percent forested (dominated by evergreens) and 15 percent herbaceous upland, shrub, and grassland (table 1). About 24 percent of the land is agricultural and 6 percent is urban in the Edisto River basin. The headwater McTier Creek subbasin has an even greater rural signature than the larger Edisto River basin, with 50 percent forested land and 22 percent herbaceous upland, shrub, and grassland lands. The McTier Creek subbasin also has less agricultural and urban land development (15 and 4.7 percent, respectively) than the Edisto River basin. The headwater McTier Creek site has reduced wetland cover of 8.2 percent as compared to 18 percent in the Edisto River basin.

Upper Hudson River Basin

The climate of this area in New York is temperate with relatively long, cold winters and a short growing season (Scudder Eikenberry and others, 2011). Mean air temperatures near the sampling sites ranged from about 18 °C in the summer to about −8 °C in the winter; annual precipitation ranged from 118.3 centimeters (cm) to 136.0 cm for the period of study (Scudder Eikenberry and others, 2011).

The Hudson River site in Essex County, New York, drains about 493 km^2 in the upper Hudson River basin (table 1). This site is in the mountainous Adirondack region, which also is part of the Northeastern Highlands Ecoregion (Omernik, 1987; U.S. Environmental Protection Agency, 2005). The upper Hudson River basin is mostly undeveloped, and some parts are protected wilderness. Land use and land cover are primarily evergreen and deciduous forests (83.7 percent), and wetlands account for a relatively high amount of land cover (9.8 percent; table 1). The Fishing Brook site is upstream of the 60-km^2 Huntington Wildlife Forest in the central Adirondack Mountains. Land use for the Fishing Brook drainage area consists primarily of upland forests (86.3 percent) and wetlands (9.3 percent). Agricultural and urban land cover represent less than 1 percent of the Fishing Brook and Upper Hudson River study basins.

Table 1. Site descriptions for selected streamflow-gaging stations in New York and South Carolina used to compute load estimates for this study.

[NAWQA, National Water-Quality Assessment Program; USGS, U.S. Geological Survey; NWIS, National Water Information System; NLCD-adjusted, National Land Cover Dataset coverages used to redefine drainage area for sites (Scudder Eikenberry and others, 2011); mi², square mile; km², square kilometer; POR, period of record; m³/s, cubic meter per second; (m³/s)/km², cubic meter per second per square kilometer; —, no data]

Type of site	USGS NWIS gaging station number	USGS NWIS station name	Site ID	NWIS drainage area		NLCD-adjusted drainage area		Latitude (decimal degrees)	Longitude (decimal degrees)	Streamflow		
				mi²	km²	mi²	km²			POR	POR mean annual stream-flow (m³/s)	POR mean unit-area streamflow [(m³/s)/km²]
Santee River NAWQA Study Unit												
Head-water stream	02172300	McTier Creek near Monetta, S.C.	McTier Creek–Monetta	15.6	40.4	15.6	40.5	33.753	−81.602	October 1995 to September 1997; February 2001 to September 2009	0.46	0.011
Head-water stream	02172305	McTier Creek near New Holland, S.C.	McTier Creek	30.7	79.5	30.7	79.4	33.718	−81.608	June 2007 to September 2009	—	—
Large river	02175000	Edisto River near Givhans, S.C.	Edisto River	2,730	7,071	2,730	7,071	33.028	−80.391	January 1939 to September 2009	70.3	0.010
Hudson River NAWQA Study Unit												
Head-water stream	0131199050	Fishing Brook (County Line Flow) near New-comb, N.Y.	Fishing Brook	25.2	65.3	25.3	65.6	43.977	−74.270	January 2007 to September 2009	—	—
Large river	01312000	Hudson River near New-comb, N.Y.	Hudson River	192	497	190	493	43.966	−74.131	October 1925 to October 1987; October 2002 to September 2009	11.6	0.024

Data Collection Methods

Streamflow and water-quality conditions in surface water and atmospheric deposition were monitored during the study at selected sites. The collected data were analyzed to identify differences among sites and to estimate loads and yields of selected constituents. Temporal extent and frequency of monitoring varied among the selected sites.

Streamflow Data Collection

Long-term (about 70 years) continuous-record USGS streamflow-gaging stations are located on the Edisto River and Hudson River at the study locations (table 1; appendix 1-A, 1-C). For the purpose of this study, the USGS also operated continuous streamflow stations at the McTier Creek and Fishing Brook sites from early to mid-2007 through September 2009 and a longer-term (about 11 years) site on McTier Creek upstream of the current study site (table 1; appendix 1-A, 1-C). Continuous streamflow data were computed at all streamflow stations using USGS standard stage/discharge techniques (Carter and Davidian, 1968; Rantz and others, 1982; Kennedy, 1984). The streamflow data are reviewed, approved, and stored in the USGS Automated Data-Processing System (ADAPS) of the National Water Information System (NWIS) database according to procedures outlined in the USGS South Carolina and New York Water Science Centers' Surface Water Quality-Assurance Plans (Cooney, 2001; Gerard Butch, New York Water Science Center, written communication, May 29, 2012). Approved surface-water data are available for retrieval on the Internet at *http://waterdata.usgs. gov/sc/nwis/sw* (South Carolina) or *http://waterdata.usgs.gov/ ny/nwis/sw* (New York).

Table 1. Site descriptions for selected streamflow-gaging stations in New York and South Carolina used to compute load estimates for this study.—Continued

[NAWQA, National Water-Quality Assessment Program; USGS, U.S. Geological Survey; NWIS, National Water Information System; NLCD-adjusted, National Land Cover Dataset coverages used to redefine drainage area for sites (Scudder Eikenberry and others, 2011); mi², square mile; km², square kilometer; POR, period of record; m³/s, cubic meter per second; (m³/s)/km², cubic meter per second per square kilometer; —, no data]

| USGS NWIS gaging station number | USGS NWIS station name | Site ID | Water quality | | 2001 NLCD land use (percent of total drainage area) | | | | | | |
			POR	Number of mercury samples	Forested	Herbaceous upland, shrub, and grassland	Agricultural	Urban	Open water	Wetlands
					Santee River NAWQA Study Unit						
02172300	McTier Creek near Monetta, S.C.	McTier Creek–Monetta	June 2007 to August 2008	8	50.0	21.0	15.4	5.3	1.0	7.3
02172305	McTier Creek near New Holland, S.C.	McTier Creek	June 2007 to August 2009	45	49.6	21.6	14.9	4.7	1.0	8.2
02175000	Edisto River near Givhans, S.C.	Edisto River	October 2005 to August 2009	25	35.9	15.2	24.3	6.0	0.5	18.1
					Hudson River NAWQA Study Unit						
0131199050	Fishing Brook (County Line Flow) near Newcomb, N.Y.	Fishing Brook	January 2007 to September 2009	41	86.3	0.70	0	0.7	3.0	9.3
01312000	Hudson River near Newcomb, N.Y.	Hudson River	August 2005 to September 2009	32	83.7	1.3	<1	0.7	4.4	9.8

Water-Quality Data Collection

Samples were collected over a range of flow conditions from January 2007 to September 2009 at the Fishing Brook site (appendix 2-C), from August 2005 to September 2009 at the Hudson River site (appendix 2-D), from June 2007 to July 2009 at the McTier Creek site (appendix 2-A), and from October 2004 to August 2009 at the Edisto River site (appendix 2-B). Forty-one stream samples were collected at Fishing Brook, and 45 stream samples were collected at McTier Creek (table 1). At the larger river basin sites, the number of collected stream samples was 25 at the Edisto River and 32 at the Hudson River (table 1). During this study period, additional water-quality and hydrologic data were collected at several locations that composed a synoptic network (appendix 2-F) within each of the study site basins and at several groundwater and surface-water locations that composed an intensive stream

reach study (appendix 2-E) within the core site basins (Scudder Eikenberry and others, 2011). The data in appendixes 2-E–2-F are provided online at *http://pubs.usgs.gov/sir/2012/5173/*, but they were not used in the analysis for this report.

Ultra-trace-level clean-sampling procedures and equipment were used to collect surface-water samples at selected sites for low-level THg and MeHg analysis (Fitzgerald and Watras, 1989; U.S. Environmental Protection Agency, 1996; Lewis and Brigham, 2004). Samples were collected at about 0.3 meter below the surface from the centroid of flow in 2-liter (L) polyethylene terephthalate (PETE) bottles. Sample processing required filtration with ultra-clean filtration devices followed by acidification of the filtrate with ultra-pure hydrochloric acid (Lewis and Brigham, 2004; Brigham and others, 2009). Aliquots of the filtrate were analyzed for FMeHg by gas chromatographic separation with cold vapor atomic fluorescence spectrometry (DeWild and others, 2002) and for

FTHg by oxidation, purge and trap, and cold vapor atomic fluorescence spectrometry (Method 1631, revision E, U.S. Environmental Protection Agency, 2002) at the USGS Mercury Research Laboratory (MRL) in Middleton, Wisconsin. The MRL also analyzed the residue on the two filters for the PMeHg and PTHg fractions using digestion followed by cold vapor atomic fluorescence spectrometry (DeWild and others, 2004; Olund and others, 2004; respectively).

Samples were collected and analyzed for nutrient, major ion, and chlorophyll concentrations by the USGS National Water Quality Laboratory (NWQL) in Denver, Colorado (U.S. Geological Survey, variously dated). Stream discharge was measured at the headwater sampling locations at the time of sample collection using depth-velocity methods measured by acoustic Doppler devices according to established USGS stream discharge methods (Rantz and others, 1982; Turnipseed and Sauer, 2010). Field properties of specific conductance, pH, dissolved oxygen, and water temperature were measured at the time of sampling with a field-calibrated multi-parameter sonde. Field property and analytical data were reviewed according to established quality-assurance and quality-control protocols and stored in the USGS NWIS database.

Additionally, samples were analyzed for total nitrogen, total phosphorus, and dissolved phases of nitrate plus nitrite, ammonia, orthophosphate, iron, manganese, silica, and major ions by the NWQL (Fishman and Friedman, 1989; Brenton and Arnett, 1993; Fishman, 1993; American Public Health Association, 1995b, 1998; Patton and Kryskalla, 2003). Samples for chlorophyll *a*, pheophytin *a*, and phytoplankton ash-free dry mass were collected on 0.47-micron glass fiber filters and analyzed by using Standard Methods and USEPA method 445.0, respectively, by the NWQL (American Public Health Association, 1995b; Arar and Collins, 1997). Total organic nitrogen and ammonia (total Kjeldahl nitrogen, or TKN) and total phosphorus (TP) concentrations were determined by analyses described by Patton and Truitt (2000) and Patton and Truitt (1992), respectively. Whole-water samples were analyzed for suspended sediment concentrations (SSCs) and sand/fine fraction at the USGS Kentucky Water Science Center Sediment Laboratory, in Louisville, Kentucky. Methods for SSCs are described in Shreve and Downs (2005).

Particulate organic carbon (POC) and particulate nitrogen (PN) concentrations were analyzed using USEPA method 440.0 by the NWQL (U.S. Geological Survey, 2000; Zimmermann and others, 1997). Analysis of dissolved organic carbon (DOC), ultraviolet absorbance at 254 nanometers (estimate of the humic content or reactive fraction of organic carbon), and organic carbon characterization was performed by the Organic Geochemistry Research Laboratory in Boulder, Colorado (Aiken and others, 1992).

Atmospheric inputs of THg concentrations and wet deposition were monitored at two National Atmospheric Deposition Program (NADP) Mercury Deposition Network (MDN) locations near each study basin: Congaree Swamp (SC19) and Huntington Wildlife Forest (NY20) (Mercury Deposition Network, 2006a; National Atmospheric Deposition Program, 2011; table 2). Precipitation samples were collected weekly and analyzed for THg using established protocols to quantify weekly precipitation and wet deposition loads of mercury at both sites (Mercury Deposition Network, 2006b; Latysh and Wetherbee, 2007). Data are available for download at *http://nadp.sws.uiuc.edu/MDN/*. Additionally, chloride and sulfate concentrations and wet deposition were monitored at two (one near each study basin) NADP National Trend Network (NTN) locations: Santee National Wildlife Refuge (SC06) and Huntington Wildlife Forest (NY20) (Mercury Deposition Network, 2006a; National Atmospheric Deposition Program, 2011; table 2).

The Parameter-Elevation Regression on Independent Slopes Model (PRISM) was used to determine basin-wide mean annual precipitation for the Edisto and Hudson River basins for this investigation. The PRISM is a system that uses point measurements of precipitation, temperature, and other climatic factors from across the United States to produce continuous, regularly spaced digital grid estimates of monthly, annual, and event-based climatic parameters (Daly and others, 1994, 2002). In a method modified from Latysh and Wetherbee (2011), PRISM grid estimates of mean annual precipitation were applied to mean annual THg concentrations at the two MDN monitoring locations to produce mean annual wet-deposition estimates for THg in both basins.

Table 2.   Selected locations in the National Atmospheric Deposition Program (NADP) Mercury Deposition Network and National Trend Network used to estimate atmospheric wet deposition of total mercury and major ions in the Edisto River basin, South Carolina, and Hudson River basin, New York.

[*http://nadp.sws.uiuc.edu/MDN/*; m, meters]

Study area	NADP monitoring location (see fig. 1)	Location name	Period of record	Latitude	Longitude	Elevation at location (m)
Mercury Deposition Network (MDN)						
Edisto River	SC19	Congaree Swamp	3/5/1996–present	33.815	−80.781	34
Hudson River	NY20	Huntington Wildlife Forest	12/10/1999–present	43.973	−74.223	500
National Trend Network (NTN)						
Edisto River	SC06	Santee National Wildlife Refuge	7/19/1994–present	33.539	−80.435	24
Hudson River	NY20	Huntington Wildlife Forest	10/31/1978–present	43.973	−74.223	500

Data Analysis Methods

Water-quality and streamflow data collected at the selected sites were analyzed using graphical and statistical techniques. The data-analysis effort focused on describing the water chemistry in the selected tributary sites and estimating loads and yields of selected filtered and particulate constituents at the McTier Creek, Edisto River, Hudson River, and Fishing Brook sites.

Land Use and Land Cover

Spatial datasets and Federal Geographic Data Committee compliant metadata for the water-sampling locations were created for the investigation and described in detail in Eikenberry and others (2011). A unique identifier was used to link the spatial data features to the analytical results for each sampling event. Public-domain datasets that were used as basemaps include digital orthophotos from 1999 and 2006, high-resolution hydrography from the National Hydrography Dataset, and 2001 land-cover data from the National Landcover Datasets (Homer and others, 2004). In addition, 1:24,000-scale USGS digital raster graphics and digital line graphs for the study area were included as base features available to users for reference. Land-use data were extracted from the 2001 National Land Cover Databases (NLCD). The NLCD is a 21-class land-cover classification scheme applied consistently over the United States (Price and others, 2007). The 2001 NLCD represents imagery and land-cover data based on Landsat 7 data from 1999 to 2003. The spatial resolution of the data is 30 meters. The 2001 NLCD definitions can be found at *http://www.mrlc.gov/nlcd01_leg.php*. The method used to group land-cover categories to determine land use is described in Fry and others (2009). Additionally, wetland coverages from the Adirondack Park Agency Wetland Effects Database (APA; New York sites) and the U.S. Fish and Wildlife Service National Wetland Inventory (NWI; South Carolina sites) were the sources of the more detailed geospatial coverage (Scudder Eikenberry and others, 2011).

Streamflow Characteristics

To obtain loading estimates with the greatest accuracy and least uncertainty, collection of water quality data should occur over several years and target a range of flow conditions. Like many constituents, Hg and DOC concentrations tend to vary with streamflow; however, that response to streamflow may vary by individual storm events, seasons, and annual climatic conditions (dry versus wet years). Large numbers of samples collected over many years that represent a range of climatic conditions (high flow, low flow, average) tend to produce more refined mean annual load estimates. In the Edisto River basin, drought conditions prevailed for water year 2008, and hydrologic conditions only partially returned to the long-term average streamflow during water year 2009. A water year is the period that extends from October 1 of one calendar year to September 30 of the next calendar year.

For this report, the term "annual mean streamflow" represents the average of daily mean streamflows over one water year at a site. The term "mean annual streamflow" represents a mean of annual mean streamflows over multiple water years based on a specified period of time. Two methods were used to evaluate flow conditions at the selected sites: (1) comparison of mean annual streamflow for the shorter-term study period to the longer-term period of record at the site or nearby long-term record site and (2) comparison of streamflow at the time of sampling to the magnitude and frequency of long-term daily mean streamflow. Annual mean streamflow for individual water years and mean annual streamflow for the period of record were obtained from published USGS annual water data reports that are available in electronic format at *http://wdr.water.usgs.gov/*. Flow duration curves (FDCs) for selected sites were used to provide the percentage of the time a certain streamflow is equaled or exceeded for a site based on the daily mean streamflows for the period of record at the site (Searcy, 1959). This method has been modified to be applied to shorter ranges of annual streamflow (Vogel and Fennessey, 1994, 1995; Castellarin and others, 2004). Streamflow data at the time of sampling at each site were overlain on the curve to verify that a fairly representative range of streamflow was sampled.

To provide load estimates over a more representative range of flow conditions, daily mean streamflows at the McTier Creek and Fishing Brook sites were extended from a beginning date of June 13, 2007, to a beginning date of October 1, 2004, based on a correlation with the continuous streamflow record at nearby index streamgaging stations. An index station is identified when streamflow data at a short-record gage are determined to be sufficiently correlated with concurrent streamflow record at a long-record gage (the index station). The index station can be used to extend the short record using record-extension techniques (Hirsch, 1982). One such method is the Maintenance of Variance Extension, Type 1 (MOVE.1) (Helsel and Hirsch, 1992). The MOVE.1 correlation maintains the mean and variance of the measured data at the short-record gage and therefore allows for the synthesis of a longer-term dataset that possesses the statistical characteristics of the measured data. Index stations used to extend streamflow were McTier Creek near Monetta (02172300) for the McTier Creek site (02172305) and Hudson River near Newcomb (01312000) for the Fishing Brook site (0131199050) (table 1).

Chemical Characteristics

Descriptive statistics including mean, maximum, minimum, and median values were graphically represented in boxplots. Median values were used for comparison because they are not strongly influenced by outliers in the data (Helsel and Hirsch, 1992). Descriptive statistics were computed using the robust regression on order statistics (ROS) method because of the presence of multiple censored threshold data (Helsel, 2005). For methylmercury, results were more frequently censored at the laboratory reporting limit (LRL) of 0.04 nanogram per liter (ng/L) for FMeHg and from 0.01 to 0.03 ng/L for PMeHg than for other constituents. The robust ROS method was implemented in Tibco Spotfire S+® 8.1 software. Box-plots serve as graphical summaries based on percentiles of the data distribution. For example, a 75th percentile represents the concentration whereby 75 percent of all concentrations measured were below that 75th percentile concentration. The 50th percentile or median represents the "middle" concentration, such that 50 percent of the data are above that concentration and 50 percent are below. The "box" displays the median 50 percent of the data (from the 75th to the 25th percentile) (Helsel and Hirsch, 1992). The "whiskers" display the data range. Outliers represent data that fall outside of 1.5 times the interquartile range.

Comparison Among Sites

To identify differences in constituent concentrations among the four sites, a non-parametric statistical analysis was performed in two steps (Helsel and Hirsch, 1992). First, a one-way analysis of variance (ANOVA) on ranked data (also called the Kruskal-Wallis test) was applied to the data to determine if a statistical difference existed among at least one of the medians of the groups of data. A Kruskal-Wallis chi-squared test statistic with a probability value below an alpha level of 0.05 was considered to denote that a statistically significant difference existed with a 95 percent confidence that the statistical finding was not false. If the more robust Kruskal-Wallis test indicated that a difference existed, then Tukey's studentized range test [a multiple comparison test also known as the honestly significantly different (HSD) test] was applied to the groups of data only to identify which group or groups were different from the others. Estimated values are defined as values below the LRL but above the method detection limit and are considered semi-quantitative. For the purposes of this report, estimated values were given the same rank above censored values and below detected values; censored values were given the same rank below estimated and detected values (Helsel, 2005).

Spearman's correlation analysis also was applied to water-quality data from each of the four sites to evaluate the strength of association among Hg species, SSC, major ion and DOC concentrations, and streamflow (Helsel and Hirsch, 1992). An alpha level of 0.05 (95 percent confidence) for significant correlations was selected for the analysis. Spearman's rho measures the observed co-variation and the strength of the monotonic relation between two variables—where change is in one direction only (either strictly rising or strictly falling, but not reversing direction). Correlation was quantified with a coefficient called "rho" for this analysis. A rho ranges from 0 to 1; the closer the rho is to 1, the stronger the correlation. A p-value is computed during the analysis. A correlation resulting in a p-value below the alpha level of 0.05 (p-value < 0.05) is considered statistically significant in this report. Caution should be expressed in interpreting correlation analysis results because a significant correlation indicates only co-variation, not cause and effect.

Load Estimation

Mercury, DOC, and SSC loads were computed using the computer program S-LOADEST. S-LOADEST is a USGS "plug-in" version for TIBCO Spotfire® S+, a commercially available statistical software package (version 8.1, TIBCO, 2008). S-LOADEST essentially replicates features of the LOAD ESTimator (LOADEST), a FORTRAN program that was developed for estimating constituent loads in streams and rivers (Runkel and others, 2004). The plug-in is programmed to allow S-LOADEST users to estimate annual, monthly, and seasonal constituent loads using statistical regression analysis incorporated in the LOADEST software program (Runkel and others, 2004) that is based on the rating curve method (Cohn and others, 1989; Cohn, 2005). The regression model computes daily loads based on relations between constituent load and one to seven explanatory variables that are functions of streamflow and time. The time component can be represented as an increasing and decreasing trend over time (decimal time term) and as seasonal changes [sine (2π decimal time) plus cosine (2π decimal time) term]. Streamflow and constituent load are transformed using the natural logarithm prior to the regression analysis to improve the fit of the regression model. To account for the bias produced in back-transformation from logarithmic to arithmetic space, the computed daily loads are adjusted by using the minimum variance unbiased estimator (for cases of no censored data) (Cohn and others, 1989) or the adjusted maximum likelihood method (AMLE) (Cohn, 2005).

Instantaneous constituent loads are computed by the following equation:

$$L_{Hg} = C_{Hg} * Q_i * C_1. \qquad (1)$$

where

L_{Hg} is the Hg species (or other constituent of interest) load at the time of sampling, in milligrams per day;

C_{Hg} is the concentration of the mercury species (or other constituent of interest), in nanograms per liter;

Q_i is the instantaneous stream discharge at the time of sampling, in cubic feet per second; and

C_1 is a unit conversion factor (2.447).

For constituents with concentrations in milligrams per liter (including DOC, SSC, and major ions), loads are computed as kilograms per day.

Yields are computed by the following equation:

$$Y_{Hg} = (L_{Hg} / DA) * C_y, \qquad (2)$$

where

Y_{Hg} is the mercury species (or other constituent of interest) yields, in micrograms per hectare per day;

L_{Hg} is the Hg species (or other constituent of
           interest) load at the time of sampling, in
           milligrams per day;

DA is the upstream basin drainage area, in
        hectares; and

C_y is a unit conversion factor (1,000).

For constituents with concentrations in milligrams per liter
(including DOC, SSC, and major ions), loads are computed as
milligrams per hectare-day.

The LOADEST program contains nine pre-defined
regression models that can be used to estimate loads that
account for the different possible combinations of explanatory
variables of streamflow and time. The model selected for this
study accounted only for streamflow and seasonality (Cohn
and others, 1989, 1992; Helsel and Hirsch, 1992). Regression
models for each constituent were examined individually based
on validation of the model. Validation checks included examination of model residuals to ensure linear fit, uniform scatter
around the fit, normality of residual distribution, and linearity
with all explanatory variables (Helsel and Hirsch, 1992). The
load equation used in this study is as follows:

$$L = \beta_0 + \beta_1 \ln Q + \beta_2 \ln Q^2 + \beta_3 \sin(2\pi T) + \beta_4 \cos(2\pi T), \quad (3)$$

where

L is the natural logarithm (ln) of the estimated
       load, in milligrams per day;

Q is the daily mean streamflow, in cubic feet
       per second;

T is centered time, in decimal years;

sin   is sine;

cos   is cosine;

π is pi; and

β_n is the estimated coefficients for each variable.

For the constituent of interest, the formulated regression
model was used to estimate loads over a selected time interval
(estimation period) of October 2004 to September 2009. Mean
load estimates, standard errors, and 95 percent confidence
intervals were developed on a seasonal and annual basis.

Litterfall Estimation

Mass of litterfall and mercury concentrations in leaf and
needle litter were estimated for the Fishing Brook and McTier
Creek headwater stream basins. Annual THg deposition from
litterfall also was computed for water year 2007 in both basins
to provide an estimate of the dry deposition of THg. In the
Fishing Brook basin, litterfall mass was estimated by placing
six 42-cm inner-diameter, open baskets in each of three forest
stands [exact location was within the tributary subbasin of
Sixmile Brook near Long Lake, N.Y. (0131199022); fig. 1].
The stand types were mixed deciduous, mixed coniferous,
and speckled alder (*Alnus incana*). The baskets passively
collected litterfall from September 11 to October 25, 2007.
In the McTier Creek basin, litterfall mass was estimated by

placing six open baskets in two forest stand types (deciduous and coniferous). The baskets passively collected litterfall
from September 7 to December 6, 2007. In both basins, all of
the leaves/needles in each basket were placed in plastic bags,
sealed, and returned to the office for drying and weighing.

Clean techniques were used to collect additional leaf
(deciduous) and needle (coniferous) litter samples for mercury
analysis. Shoulder-length gloves were used when collecting
litter samples for mercury analysis. These samples represented leaf litter collected near the litterfall baskets (combined
deciduous and coniferous) that had not been in contact with
the baskets. The litter was picked up off the ground immedi-
ately after it fell (to minimize post-deposition changes in Hg
concentrations) during one day in the middle of the litterfall
season at each site. Leaf-litter samples were freeze-dried
overnight. After thawing and drying, samples were pulverized
to a powder with a pre-cleaned mortar and pestle and placed
in a glass tube for shipment to the USGS MRL for THg and
MeHg analysis.

For Fishing Brook, 1992 NLCD, National Wetland
Inventory, and Adirondack Park Agency geographic information system data were used to extrapolate the litterfall estimates
to the basin scale, based on type of forest coverage (mixed,
deciduous, coniferous categories). Assumptions were made that
mixed forests consisted of 50 percent deciduous and 50 percent
coniferous and that the litterfall occurred predominantly in the
fall. Litterfall rates were computed as grams of leaf litter mass
per square meter per year at each site and were used to calcu-
late annual THg and MeHg deposition rates, in micrograms
per square meter per year, by multiplying the Hg concentration
in the leaf litter by the litterfall rate for each stand type. Then
THg and MeHg deposition rates were extrapolated over the
total stand type area for the basin to obtain the estimated total
basin-wide annual THg and MeHg deposition, in grams.

Quality-Assurance and Quality-Control Data

Of the samples collected at the McTier Creek site
(02172305), 17 percent were collected for quality-assurance,
quality-control (QAQC) purposes. The QAQC samples
included seven field blanks for Hg and DOC species related to
all aspects of field collection procedures. FTHg concentrations
in field blanks ranged from 0.07 to 0.29 ng/L and averaged
0.13 ng/L (±0.07 standard deviation). These levels were considered negligible because they represented less than 5 percent
on average of the environmental sample concentrations.
PTHg, PMeHg, and FMeHg concentrations in field blanks
consistently were below the LRL. DOC concentrations in field
blanks ranged from 0.2 to 0.7 milligrams per liter (mg/L) and
averaged 0.4 mg/L (±0.2 standard deviation). These levels
were considered negligible because they represented less than
7 percent, on average, of the environmental sample concentrations. Relative percent difference (RPD), expressed in percent,

between the concentration in the replicate sample (C_{rep}) and the concentration in the environmental sample (C_{env}) was computed by subtracting C_{rep} from C_{env}, dividing that difference by the average of C_{rep} and C_{env}, and multiplying by 100. A replicate was collected at average to high-flow conditions [0.85 cubic meter per second (m^3/s)] at the McTier Creek site. This replicate had RPDs of 30 and 0 percent for PTHg and PMeHg, respectively; however, the RPD for THg was reduced to 10 percent, and MeHg remained at 0 percent because RPDs for FTHg and FMeHg were 0 percent.

Of the samples collected at Fishing Brook (0131199050), 17 percent were collected for QAQC purposes. The QAQC samples included five field blanks for Hg and DOC species related to all aspects of field collection procedures. FTHg concentrations ranged from less than 0.04 to 0.08 ng/L and averaged 0.06 ng/L (±0.02 standard deviation). These levels were considered negligible because they represented less than 3 percent, on average, of the environmental sample concentrations. PTHg, PMeHg, and FMeHg concentrations in field blanks consistently were below the LRLs. DOC concentrations in field blanks ranged from 0.4 to 0.9 mg/L and averaged 0.5 mg/L (±0.2 standard deviation). These levels were considered negligible because they represented less than 7 percent, on average, of the environmental sample concentrations. Two replicates were collected at the Fishing Brook site: one under low-flow conditions (0.11 m^3/s) and one under high-flow conditions (3.3 m^3/s). During low-flow conditions, RPDs for FTHg and FMeHg were 1.5 and 10 percent, respectively. During high-flow conditions, RPDs increased to 14 and 28 percent for FTHg and FMeHg, respectively. A similar pattern of increasing RPDs at high flow was observed for DOC and particulate species of Hg.

Fluvial Transport of Selected Constituents

In this section, streamflow and chemical characteristics are described and compared among the McTier Creek, Edisto River, Fishing Brook, and Hudson River sites. The relations of streamflow to mercury species, DOC, and SSC have been evaluated and are compared among the four sites. Estimates of mercury species, DOC, SSC, and major ion loads and yields are described and compared among the four sites. Estimated stream yields of THg, chloride, and sulfate are compared to estimates of atmospheric wet deposition and litterfall deposition (mercury only) at the McTier Creek and Fishing Brook sites.

Streamflow Characteristics

Annual mean streamflow and annual mean unit-area streamflow were computed for each site to help characterize flow conditions for the study period that includes water years 2005 to 2009 (fig. 2; table 3; appendix 1-A). Mean annual

streamflow for the period of record at the Hudson River (01312000), Edisto River (02175000), and McTier Creek–Monetta (02172300) also was computed (fig. 2; table 1). McTier Creek (02172305) and Fishing Brook (0131199050) had only 2 water years with complete streamflow record (2008 and 2009), so mean annual streamflow was not computed for those two sites. McTier Creek had annual mean streamflow of 0.54 and 0.75 cubic meters per second (m^3/s) for 2008 and 2009, respectively (fig. 2; table 3). Fishing Brook had annual mean streamflows of 1.75 and 1.58 m^3/s, respectively, for the same 2-year period (fig. 2; table 3). During the 2005 to 2009 study period, annual mean streamflow ranged from 11.7 to 17.1 m^3/s at the Hudson River site (fig. 2; table 3). For the Hudson River, annual mean streamflow for the study period was consistently above the long-term (1925 to 2009) mean annual streamflow of 11.6 m^3/s (fig. 2; table 3), indicating relatively high-flow conditions. Annual mean streamflow ranged from 26.3 to 50.0 m^3/s at the Edisto River site for the 5-year study period (fig. 2; table 3). For the Edisto River site, annual mean streamflow for the study period was consistently below the long-term (1940 to 2009) mean annual streamflow of 70.3 m^3/s, indicating low-flow conditions. Because limited streamflow record existed for the McTier Creek site, annual mean streamflow for the study period at the McTier Creek–Monetta site was compared to the long-term (1996 to 2009) mean annual streamflow of 0.46 m^3/s to evaluate flow conditions during the study period (table 1). Annual mean streamflow at McTier Creek–Monetta ranged from 0.28 to 0.52 m^3/s and was consistently below the mean annual streamflow for this site except for the annual mean streamflow for 2005 (0.52 m^3/s; fig. 2; table 3).

Dividing the annual mean streamflow by drainage area computes an annual mean unit-area streamflow, in cubic meters per second per square kilometers. This expression allowed comparison of streamflow among sites of different basin areas (fig. 3; table 3). For the period of study, sites in the Hudson River basin had an order of magnitude higher streamflow per unit area than sites in the Edisto River basin, which further illustrates the differences in flow conditions between the two basins. Differences within basins also were observed. The larger-scale Edisto River site consistently had lower annual mean unit-area streamflow than the two headwater stream sites in McTier Creek (fig. 3). The opposite pattern was observed in the Hudson River basin, where the larger Hudson River site had lower annual mean unit-area streamflow than the Fishing Brook site (fig. 3).

Flow duration curves (FDCs) provided further characterization of streamflow (Searcy, 1959; Vogel and Fennessey, 1994, 1995). The FDCs at McTier Creek–Monetta, Edisto River, and Hudson River sites, which had long-term periods of streamflow record, provided more steady-state-based information on the streamflow exceedances than did the FDCs at the McTier Creek and Fishing Brook sites, which had streamflow record only for the study period (partial water year 2007 to 2009) (figs. 4, 5; table 1). Because the interpretation of an FDC depends upon the period of record over which the FDC

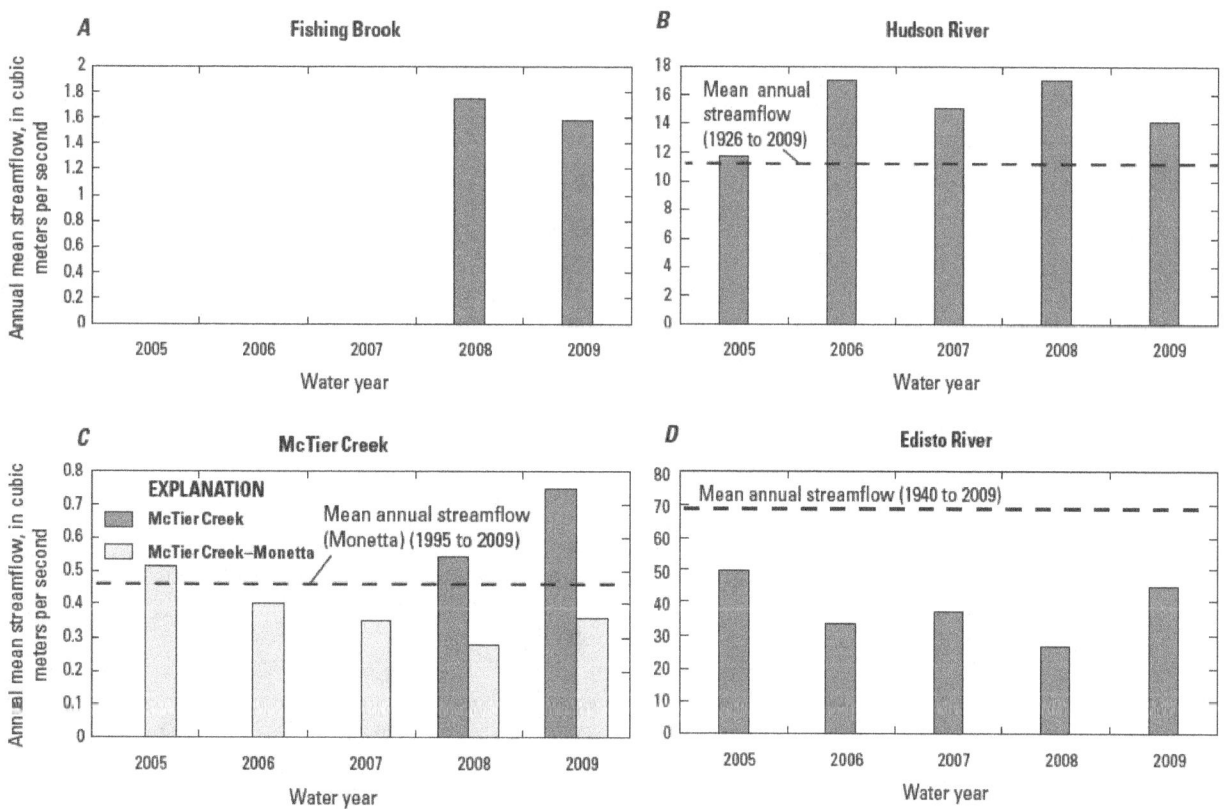

Figure 2. Annual mean streamflow for the study period at (*A*) Fishing Brook at County Line Flow near Newcomb, N.Y. (station 0131199050), (*B*) Hudson River near Newcomb, N.Y. (station 01312000), (*C*) McTier Creek near New Holland, S.C. (McTier Creek, station 02172305) and McTier Creek near Monetta, S.C. (McTier Creek–Monetta, station 02172300), and (*D*) Edisto River near Givhans, S.C. (station 02175000). [m³/s, cubic meters per second; WY, water year]

Table 3. Annual mean streamflow and unit-area streamflow at three sites in the Edisto River basin in South Carolina (McTier Creek and Edisto River) and at two sites in the upper Hudson River basin in New York (Fishing Brook and Hudson River) for water years 2005 to 2009.

[ND, no data; WY, water year]

Station number	Site (see table 1 for site information)	Annual mean streamflow, in cubic meters per second (cubic feet per second)					Annual mean unit-area streamflow, in cubic meters per square kilometer per second (cubic feet per square mile per second)				
		WY 2005	WY 2006	WY 2007	WY 2008	WY 2009	WY 2005	WY 2006	WY 2007	WY 2008	WY 2009
02172300	McTier Creek–Monetta	0.52 (18.2)	0.40 (14.2)	0.35 (12.5)	0.28 (9.94)	0.36 (12.7)	0.013 (1.17)	0.010 (0.910)	0.0088 (0.801)	0.0070 (0.637)	0.0089 (0.814)
02172305	McTier Creek	ND	ND	ND	0.54 (19.2)	0.75 (26.4)	ND	ND	ND	0.0068 (0.625)	0.0094 (0.860)
02175000	Edisto River	50.0 (1,765)	33.8 (1,193)	37.0 (1,306)	26.3 (929.4)	44.7 (1,579)	0.0071 (0.647)	0.0048 (0.437)	0.0052 (0.478)	0.0037 (0.340)	0.0063 (0.578)
0131199050	Fishing Brook	ND	ND	ND	1.75 (61.8)	1.58 (55.9)	ND	ND	ND	0.027 (2.45)	0.024 (2.22)
01312000	Hudson River	11.7 (413.2)	17.1 (603.5)	15.1 (532.2)	17.1 (603.2)	14.1 (498.1)	0.024 (2.15)	0.0344 (3.14)	0.0304 (2.77)	0.034 (2.14)	0.028 (2.59)

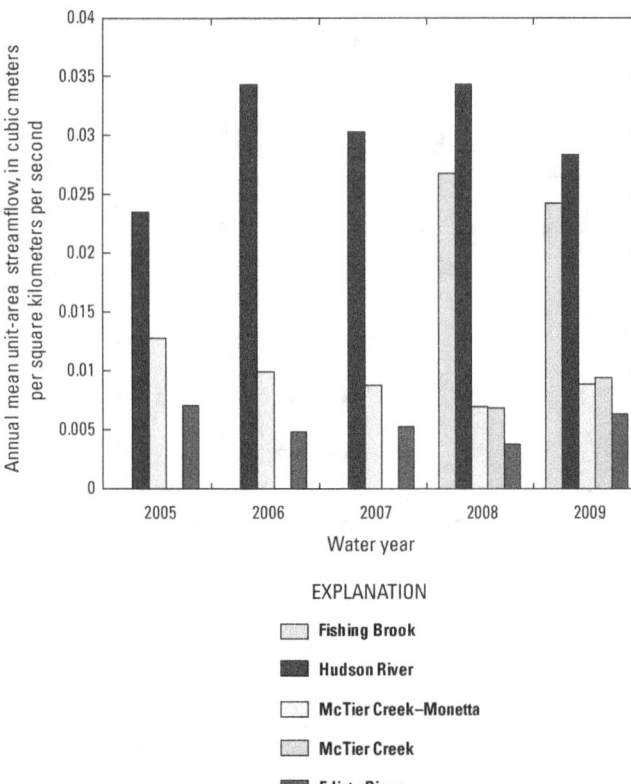

Figure 3. Annual mean unit-area streamflow at Fishing Brook at County Line Flow near Newcomb, N.Y. (Fishing Brook; station 0131199050), Hudson River near Newcomb, N.Y. (Hudson River; station 01312000), McTier Creek near Monetta, S.C. (McTier Creek–Monetta, station 02172300), McTier Creek near New Holland, S.C. (McTier Creek, station 02172305) and Edisto River near Givhans, S.C. (Edisto River; station 02175000).

was computed, only the long-term FDCs should be compared among sites. The FDCs for the Edisto and Hudson River sites have relatively flat slopes, indicating that streamflow is well sustained by surface releases or groundwater discharge (figs. 4, 5). However, the Edisto River site has a slightly more sustained low-flow characteristic than the Hudson River site, based on the lower ends of the FDCs (figs. 4, 5).

Measured streamflow at the time of water-quality sampling (contemporaneous) was plotted on the FDC for each site to determine the statistical nature of the contemporaneous streamflow as compared to historical data and thus identify limitations or strengths to the load estimation model due to unrepresentative or representative streamflow conditions. For the Edisto River site, higher streamflow conditions [greater than 113.3 m^3/s (4,000 ft^3/s) or less than 20 percent exceedance] were not well represented by the sampling. It should be noted that the FDC represented a 30-year period that would be indicative of the full range of flow conditions, but as demonstrated earlier, the 5-year estimation period for the load model was dominated by low-flow conditions. In fact, the maximum daily mean streamflow at the Edisto River site covered by the

load estimation model (water years 2005 to 2009) was within that range [154.4 m^3/s (5,454 ft^3/s) or about 15 percent exceedance] (appendix 1-C).

Daily mean streamflows at the McTier Creek (02172305) and Fishing Brook (0131199050) sites were extended to October 2004 based on correlations with the continuous streamflow record at nearby long-term gaging stations and using the MOVE.1 method. The Pearson correlation coefficient between mean daily streamflow at the McTier Creek station (02172305) and the long-term McTier Creek near Monetta station (02172300) was 0.95. The MOVE.1 estimates for McTier Creek for the period of June 13, 2007, to September 30, 2009, had a root mean square error (RMSE) of 0.23 m^3/s (8.13 ft^3/s), a coefficient of determination [R-squared (R^2)] of 0.85, and a Nash-Sutcliffe index of 0.83. The Nash-Sutcliffe index or model efficiency coefficient is commonly used as one measure to assess the performance of hydrologic models, whereby a perfect match of modeled discharge to the observed data would correspond to an index of 1 (Nash and Sutcliffe, 1970; Jain and Sudheer, 2008). Comparison of the concurrent measured streamflows at Fishing Brook (0131199050) and Hudson River (01312000) produced the Pearson correlation coefficient of 0.93 that indicated that the Hudson River station explained about 93 percent of the variability in the data at Fishing Brook. The diagnostics for the MOVE.1 estimates for Fishing Brook for the period of January 25, 2007, to September 30, 2009, were the RSME of 0.78 m^3/s (27.4 ft^3/s), R^2 of 0.82, and Nash-Sutcliffe index of 0.91.

Chemical Characteristics of the Study Sites

Basic water chemistry characteristics and organic carbon, suspended sediment, and mercury species concentrations were summarized statistically and compared to determine if differences existed among the four sites in the Edisto and upper Hudson River basins (table 4; appendixes 2-A–2-D). Dissolved iron, sulfate, calcium, and chloride concentrations were selected to represent the basic water chemistry of the sites (fig. 6).

Among the four sites, median dissolved iron concentrations ranged from 101 micrograms per liter (µg/L) at the Hudson River site to 657 µg/L at the McTier Creek site (table 4). Median dissolved sulfate concentrations ranged from 1.20 mg/L (McTier Creek site) to 7.44 mg/L (Edisto River site). A large range in median sulfate was apparent within the upper Hudson River basin, where the median dissolved sulfate concentration of 5.86 mg/L at the Hudson River site was almost twice the median of 3.44 mg/L at the headwater stream site of Fishing Brook. As with median dissolved sulfate concentrations, McTier Creek had the lowest median dissolved calcium and chloride concentrations of 0.77 and 3.15 mg/L, respectively, and the Edisto River had the highest median dissolved calcium and chloride concentrations of 6.40 and 9.88 mg/L, respectively.

Results from Kruskal-Wallis and Tukey's tests identified significant differences in basic water chemistry between

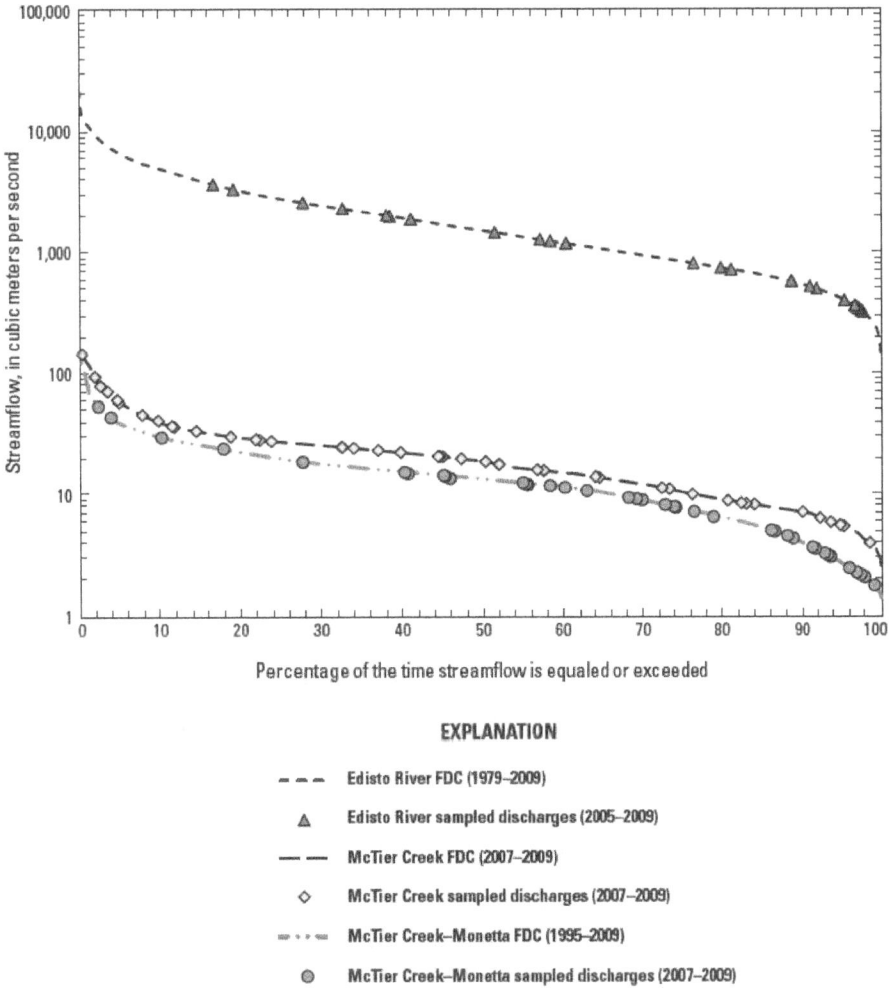

Streamflow, in cubic meters per second

Percentage of the time streamflow is equaled or exceeded

EXPLANATION

- - - Edisto River FDC (1979–2009)

△ Edisto River sampled discharges (2005–2009)

— — McTier Creek FDC (2007–2009)

◇ McTier Creek sampled discharges (2007–2009)

— · · — McTier Creek–Monetta FDC (1995–2009)

◉ McTier Creek–Monetta sampled discharges (2007–2009)

Figure 4. Flow duration curves (FDC) and sampled streamflow at McTier Creek near New Holland, S.C. (station number 02172305), McTier Creek near Monetta, S.C. (station number 02172300), and Edisto River near Givhans, S.C. (station number 02175000) for variable periods of record.

the paired headwater stream and larger river sites within the Edisto and Hudson River basins and significant differences between the two basins (p-value < 0.001; fig. 6). McTier Creek, the headwater stream in the Edisto River basin, had significantly greater dissolved iron concentrations and significantly less dissolved calcium and sulfate concentrations than the Edisto River, Hudson River, and Fishing Brook sites (p-value < 0.001; fig. 6). Similar patterns in differences in dissolved iron and sulfate concentrations were identified between the headwater stream site, Fishing Brook, and the larger Hudson River site. The Edisto River had the greatest sulfate, calcium, and chloride concentrations of the four sites, and the Hudson River had the least dissolved iron concentrations of the four sites (p-value < 0.001; fig. 6).

Among the four sites, median DOC concentrations ranged from 5.4 mg/L at the Hudson River site to 8.5 mg/L at the Edisto River site (table 4). The particulate fraction of organic carbon (POC) had median concentrations that were an order of magnitude lower than the median concentrations of the dissolved fraction (DOC). Median POC concentrations

ranged from 0.18 mg/L at the Hudson River site to 0.93 mg/L at the McTier Creek site (table 4).

Concentrations of DOC and POC in water were compared among the four sites (fig. 7; table 4). The larger Edisto River site had greater DOC concentrations than its paired headwater stream site, McTier Creek, and the two sites in the upper Hudson River basin (p-value < 0.001; fig. 7). However, Fishing Brook, the headwater stream site for the Hudson River basin, had similar DOC and POC concentrations, compared to those of the larger Hudson River site. Between river basins, the McTier Creek site had greater POC concentrations than sites in the upper Hudson River basin.

SSCs represented the inorganic and organic fractions of suspended (particulate) material in the water column at the time of sampling. Median SSCs ranged from 1 mg/L at the Hudson River site to 6 mg/L at both the McTier Creek and Edisto River sites (table 4). Additionally, medians of the percentage of the suspended sediment that was finer than 63 microns (representative of silt- and clay-sized sediment) ranged from 80 percent at the Edisto River site to 95 percent at

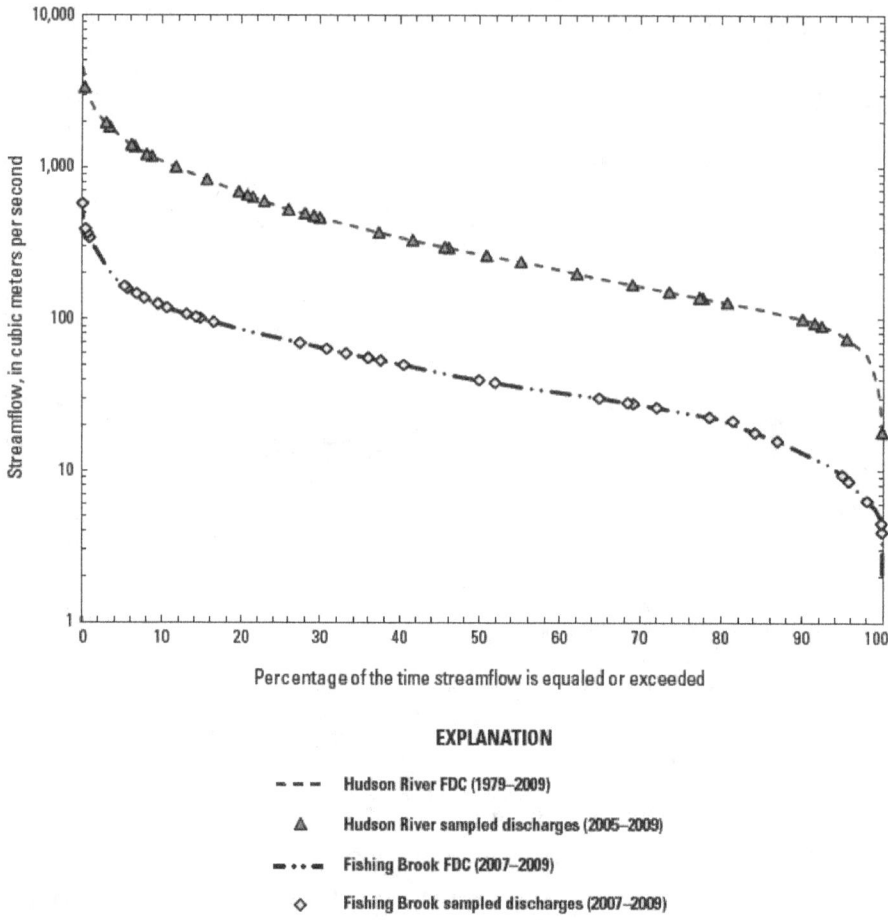

EXPLANATION

- - -   Hudson River FDC (1979–2009)

▲   Hudson River sampled discharges (2005–2009)

- ·· -   Fishing Brook FDC (2007–2009)

◇   Fishing Brook sampled discharges (2007–2009)

Figure 5.   Flow duration curves (FDC) and sampled streamflow at Fishing Brook at County Line Flow near Newcomb, N.Y. (station number 0131199050) and Hudson River near Newcomb, N.Y. (station number 01312000) for the study period (2007 to 2009) and the historical 30-year period of record, respectively.

the Fishing Brook site. The minimum percentage of silt- and clay-sized particles at the paired sites in the upper Hudson River basin remained above 50 percent for all collected samples, whereas minimum percentages at McTier Creek and Edisto River sites fell well below 50 percent (38 and 6 percent, respectively) indicating that, at least periodically, sand-sized sediment dominated the SSC.

The distribution of SSCs followed a similar pattern to POC concentrations among sites in both basins (fig. 7). Within the Edisto and Hudson River basins, no differences in SSCs were identified between the paired headwater stream and river sites (p-value < 0.001). Between river basins, the Edisto River site had greater SSCs than sites in the upper Hudson River basin (p-value < 0.001). Suspended sediment at the Fishing Brook site had a greater percentage of silt- and clay-sized particles than suspended sediment at the McTier Creek and Edisto River sites (p-value < 0. 001). The presence of 325 small run-of-river lakes along the upper Hudson River (including County Line Flow at the Fishing Brook site) could contribute

to the statistically lower concentrations of particulate material when compared to the Edisto River.

Median concentrations of Hg species at sites in the Edisto River basin were consistently greater than those at sites in the Hudson River basin. Median concentrations of FTHg ranged from 1.55 ng/L at the Hudson River site to 2.77 ng/L at the Edisto River site (table 4). The Edisto River site had the greatest median FMeHg concentration of 0.32 ng/L, which was at least 3 times higher than concentrations at the other three sites, and the Hudson River site had the least median FMeHg concentration of 0.07 ng/L. The McTier Creek and Edisto River sites had median PTHg concentrations of 0.99 and 1.19 ng/L, respectively, which were more than 2 times larger than the concentrations at the Fishing Brook and Hudson River sites (0.40 and 0.30 ng/L, respectively). Median PMeHg concentrations ranged from 0.01 ng/L at the Hudson River site to 0.05 ng/L at the Edisto River site.

Differences in filtered and particulate Hg concentrations were identified among sites. Concentrations of FTHg were

Table 4.   Summary statistics for selected constituent concentrations at McTier Creek near New Holland, S.C., Edisto River near Givhans, S.C., Fishing Brook at County Line Flow near Newcomb, N.Y., and Hudson River near Newcomb, N.Y., 2005 to 2009.

[n (censored), number of samples (number of censored results); StDev, standard deviation; 25th Q, twenty-fifth quartile; 75th Q, seventy-fifth quartile; C, degrees Celsius; µS/cm, microsiemens per centimeter at 25 degrees Celsius; ng/L, nanograms per liter; <, less than the laboratory reporting level; min, minimum; max, maximum; mg/L, milligrams per liter; µg/L, micrograms per liter; red highlighted text, used regression on order statistics (ROS) for determining descriptive statistics for constituents with censored values]

Constituent	n (censored)	Units	Mean	StDev	Median	25th Q	75th Q	Min	Max
			McTier Creek near New Holland, S.C. (02172305)						
Water temperature	44 (0)	°C	16.2	6.8	18.4	9.5	22.0	3.0	25.7
pH	44 (0)	standard units	5.0	0.6	5.0	4.7	5.4	3.3	5.9
Specific conductance	45 (0)	µS/cm	22.5	2.3	22.0	21.0	24.0	19.0	31.0
Dissolved oxygen	42 (0)	mg/L	8.6	2.2	7.8	6.9	10.2	5.1	13.6
Filtered total mercury	43 (0)	ng/L	2.68	1.7	2.16	1.74	2.98	0.82	9.88
Filtered methylmercury	45 (4)	ng/L	0.11	0.06	0.11	0.08	0.16	<0.04	0.24
Particulate total mercury	45 (1)	ng/L	1.34	0.99	0.99	0.78	1.43	<0.11	5.28
Particulate methylmercury	45 (13)	ng/L	0.04	0.04	0.03	<0.01	0.05	<0.01	0.27
Suspended sediment	44 (0)	mg/L	10	12	6	4	9	1	58
Suspended sediment finer than 63 microns	38 (0)	percent	77	13	81	71	86	38	92
Particulate organic carbon	40 (0)	mg/L	1.6	1.6	0.93	0.65	1.7	0.38	8
Dissolved organic carbon	44 (0)	mg/L	7.0	2.1	6.4	5.6	8.0	4.3	14
Dissolved iron	45 (0)	µg/L	642	253	657	443	790	230	1,159
Dissolved sulfate	45 (0)	mg/L	1.25	0.70	1.20	0.64	1.77	0.35	2.78
Dissolved calcium	45 (0)	mg/L	0.78	0.07	0.77	0.74	0.82	0.63	0.95
Dissolved chloride	45 (0)	mg/L	3.16	0.26	3.15	2.99	3.29	2.44	3.85
			Edisto River near Givhans, S.C. (02175000)						
Water temperature	50 (0)	°C	19.5	7.0	20.2	13.7	26.4	7.1	28.9
pH	50 (0)	standard units	6.3	0.5	6.4	6.0	6.6	5.0	7.0
Specific conductance	50 (0)	µS/cm	90.8	17.3	86.5	80.3	101	55.0	137.0
Dissolved oxygen	50 (0)	mg/L	7.9	1.6	7.4	6.6	9.5	5.2	11.1
Filtered total mercury	23 (0)	ng/L	3.27	1.79	2.77	1.85	4.29	1.17	7.75
Filtered methylmercury	24 (0)	ng/L	0.31	0.16	0.32	0.17	0.37	0.04	0.69
Particulate total mercury	24 (1)	ng/L	1.41	1.06	1.19	0.71	1.94	<0.05	5.37
Particulate methylmercury	24 (1)	ng/L	0.06	0.07	0.05	0.02	0.07	<0.01	0.35
Suspended sediment	50 (0)	mg/L	11	24	6	4	9	1	169
Suspended sediment finer than 63 microns		percent	75	19	80	71	85	6	97
Particulate organic carbon	10 (0)	mg/L	0.76	0.44	0.72	0.44	0.9	0.23	1.7
Dissolved organic carbon	24 (0)	mg/L	8.8	4.3	8.5	4.9	11.1	3.4	20.1
Dissolved iron	31 (0)	µg/L	307	139	289	227	388	78.7	703
Dissolved sulfate	49 (0)	mg/L	8.73	3.50	7.44	5.94	11.8	3.33	18.2
Dissolved calcium	31 (0)	mg/L	6.16	1.57	6.40	5.05	7.08	3.11	10.4
Dissolved chloride	49 (0)	mg/L	10.2	2.5	9.88	8.61	11.5	6.12	19.6

Table 4. Summary statistics for selected constituent concentrations at McTier Creek near New Holland, S.C., Edisto River near Givhans, S.C., Fishing Brook at County Line Flow near Newcomb, N.Y., and Hudson River near Newcomb, N.Y., 2005 to 2009.—Continued

[n (censored), number of samples (number of censored results); StDev, standard deviation; 25th Q, twenty-fifth quartile; 75th Q, seventy-fifth quartile; C, degrees Celsius; µS/cm, microsiemens per centimeter at 25 degrees Celsius; ng/L, nanograms per liter; <, less than the laboratory reporting level; min, minimum; max, maximum; mg/L, milligrams per liter; µg/L, micrograms per liter; red highlighted text, used regression on order statistics (ROS) for determining descriptive statistics for constituents with censored values]

Constituent	n (censored)	Units	Mean	StDev	Median	25th Q	75th Q	Min	Max
Fishing Brook at County Line Flow near Newcomb, N.Y. (0131199050)									
Water temperature	43 (0)	°C	9.9	8.5	9.6	1.0	18.2	–0.1	23.8
pH	43 (0)	standard units	6.3	0.5	6.3	5.9	6.7	5.6	8.0
Specific conductance	43 (0)	µS/cm	47.0	12.0	48.0	38.1	54.2	18.4	69.0
Dissolved oxygen	43 (0)	mg/L	10.1	2.5	10.1	7.7	11.9	5.8	15.2
Filtered total mercury	41 (0)	ng/L	2.02	0.61	2.05	1.52	2.4	1.04	3.42
Filtered methylmercury	41 (6)	ng/L	0.14	0.10	0.10	0.07	0.18	<0.04	0.45
Particulate total mercury	41 (0)	ng/L	0.40	0.16	0.40	0.27	0.49	0.13	0.69
Particulate methylmercury	41 (8)	ng/L	0.03	0.02	0.02	0.01	0.04	<0.01	0.07
Suspended sediment	44 (0)	mg/L	3	2	2	1	3	1	12
Suspended sediment finer than 63 micron	30 (0)	percent	93	6	95	89	99	78	100
Particulate organic carbon	41 (5)	mg/L	0.43	0.32	0.30	0.14	0.73	<0.12	1.1
Dissolved organic carbon	43 (0)	mg/L	7.1	2.0	7.0	5.8	8.4	3.9	11.7
Dissolved iron	40 (0)	µg/L	310	200	228	156	463	81.5	810
Dissolved sulfate	40 (0)	mg/L	3.56	0.77	3.44	2.96	4.04	2.42	5.34
Dissolved calcium	40 (0)	mg/L	3.99	1.05	4.12	3.23	4.67	2.11	6.01
Dissolved chloride	40 (0)	mg/L	5.27	1.64	5.11	4.08	6.28	1.96	8.33
Hudson River near Newcomb, N.Y. (01312000)									
Water temperature	34 (0)	°C	10.1	8.3	9.9	1.3	18.1	0	23.0
pH	34 (0)	standard units	6.6	0.4	6.7	6.4	6.9	5.6	7.3
Specific conductance	34 (0)	µS/cm	45.0	12.4	41.7	36.8	52.8	20.2	72.0
Dissolved oxygen	34 (0)	mg/L	10.7	2.3	10.9	8.7	12.7	6.4	14.8
Filtered total mercury	32 (0)	ng/L	1.59	0.3	1.55	1.35	1.7	1.14	2.40
Filtered methylmercury	32 (4)	ng/L	0.09	0.06	0.07	0.04	0.12	<0.04	0.24
Particulate total mercury	32 (0)	ng/L	0.40	0.49	0.30	0.23	0.36	0.17	2.98
Particulate methylmercury	32 (9)	ng/L	0.02	0.01	0.01	0.01	0.02	<0.01	0.06
Suspended sediment	34 (3)	mg/L	2	2	1	1	2	<0.5	13
Suspended sediment finer than 63 microns	12 (0)	percent	83	16	89	68	97	59	100
Particulate organic carbon	32 (5)	mg/L	0.22	0.16	0.18	0.14	0.25	<0.12	0.89
Dissolved organic carbon	35 (0)	mg/L	5.7	1.7	5.4	4.7	6.2	3.7	13.4
Dissolved iron	32 (0)	µg/L	104	36.5	101	75.7	129	49.2	178
Dissolved sulfate	32 (0)	mg/L	6.22	1.94	5.86	4.64	7.07	3.66	11
Dissolved calcium	32 (0)	mg/L	4.46	1.19	4.2	3.7	5.07	2.52	7.31
Dissolved chloride	32 (0)	mg/L	2.82	0.77	2.83	2.34	3.09	1.29	5.32

less at the paired sites in the upper Hudson River basin than at the Edisto River site (p-value < 0.001; fig. 8). Additionally, the Hudson River site had significantly lower FTHg concentrations than the two headwater catchment sites (McTier Creek and Fishing Brook). The Edisto River site had significantly greater FMeHg concentrations than the McTier Creek, Fishing Brook, and Hudson River sites (p-value < 0.001). Differences in the particulate concentrations of Hg were similar to the differences identified among sites for SSCs and POC concentrations. Sites in the Edisto River basin had greater PTHg concentrations than sites in the upper Hudson River basin (p-value < 0.001). Although the headwater sites had similar PMeHg concentrations, the Edisto River site had significantly higher PMeHg concentrations than the Fishing Brook and Hudson River sites (p-value < 0.001).

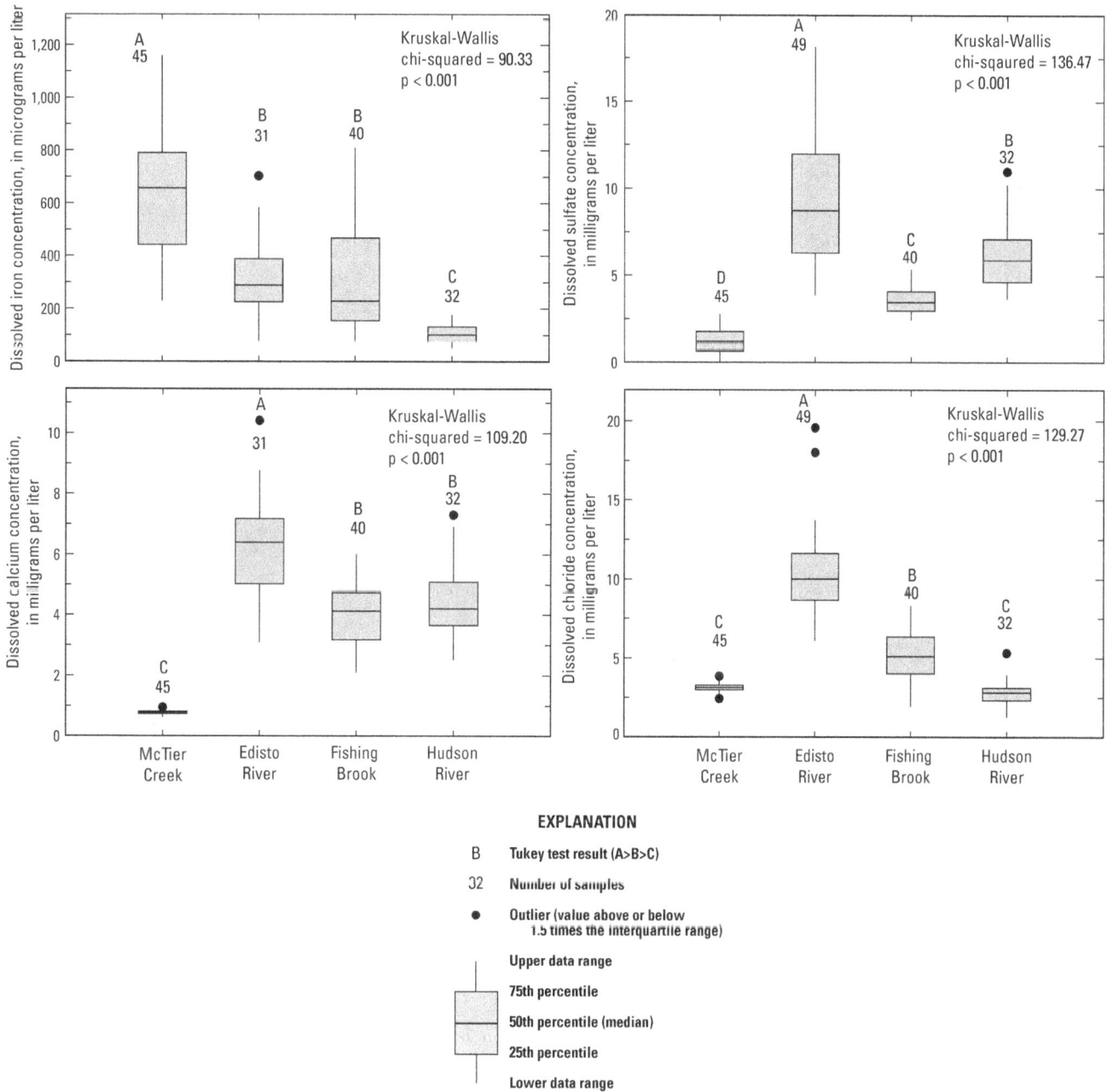

Figure 6. Distributions of dissolved iron concentrations, dissolved sulfate concentrations, dissolved calcium concentrations, and dissolved chloride concentrations for the McTier Creek near New Holland, S.C. (McTier Creek, station 02172305), Edisto River near Givhans, S.C. (Edisto River, station 02175000), Fishing Brook at County Line Flow near Newcomb, N.Y. (Fishing Brook, station 0131199050), and Hudson River near Newcomb, N.Y. (Hudson River, station 01312000) sites, 2007 to 2009.

Relation of Streamflow to Mercury Species, Dissolved Organic Carbon, and Suspended Sediment

At the McTier Creek site, particulate and filtered concentrations of THg were positively correlated with streamflow, indicating an increased watershed contribution during high streamflow conditions (table 5; FTHg p-value < 0.001; PTHg p-value = 0.012). Direct atmospheric wet deposition, terrestrial runoff, and increased wetland connection to the stream could serve as processes that deliver THg to the stream during high streamflow conditions. However, FMeHg concentrations at this site were negatively correlated with streamflow, indicating a dilution by FMeHg-depleted waters during high streamflow periods (FMeHg p-value < 0.001). Streamflow did not

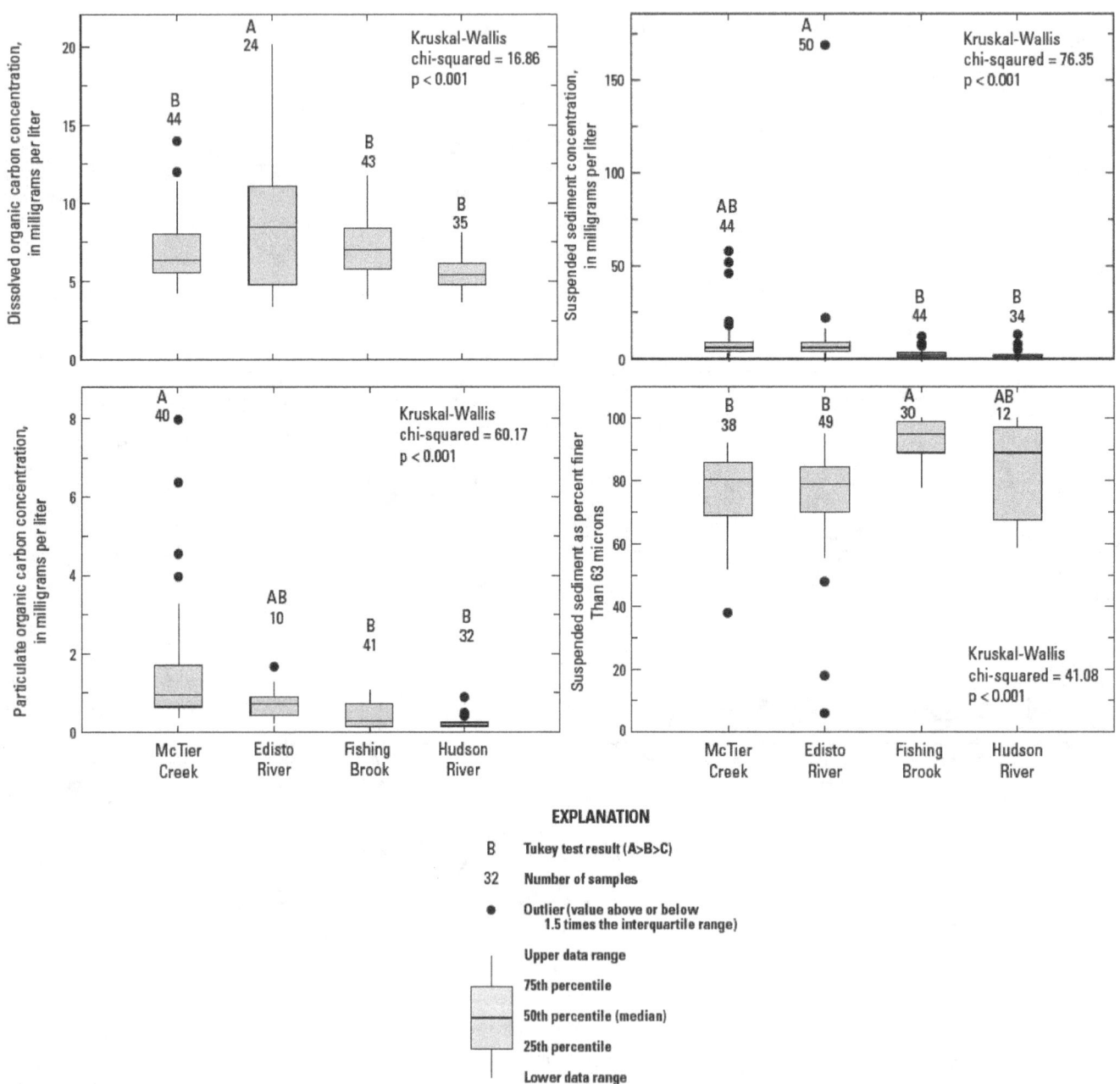

Figure 7. Distributions of dissolved organic carbon concentrations, suspended sediment concentrations, particulate organic carbon concentrations, and percentage of suspended sediment finer than 63 microns for the McTier Creek near New Holland, S.C. (McTier Creek, station 02172305), Edisto River near Givhans, S.C. (Edisto River, station 02175000), Fishing Brook at County Line Flow near Newcomb, N.Y. (Fishing Brook, station 0131199050), and Hudson River near Newcomb, N.Y. (Hudson, station 01312000) sites, 2007 to 2009.

co-vary with PMeHg concentrations at the McTier Creek site. At the Edisto River site, FTHg and FMeHg concentrations were correlated positively with streamflow, which is indicative of increased watershed contribution during high streamflow conditions (FTHg p-value < 0.001; FMeHg p-value = 0.012); however, PTHg and PMeHg concentrations did not co-vary significantly with streamflow. At the McTier Creek and Edisto River sites, streamflow and DOC concentrations had

statistically significant positive correlations similar to those for FTHg. As was observed with the particulate forms of Hg, streamflow was not correlated significantly to SSC concentrations at the McTier Creek and Edisto River sites (table 5).

The filtered forms of mercury at the Fishing Brook site exhibited a similar pattern to those at the McTier Creek site, whereby FTHg concentrations were positively correlated and FMeHg concentrations were negatively correlated to

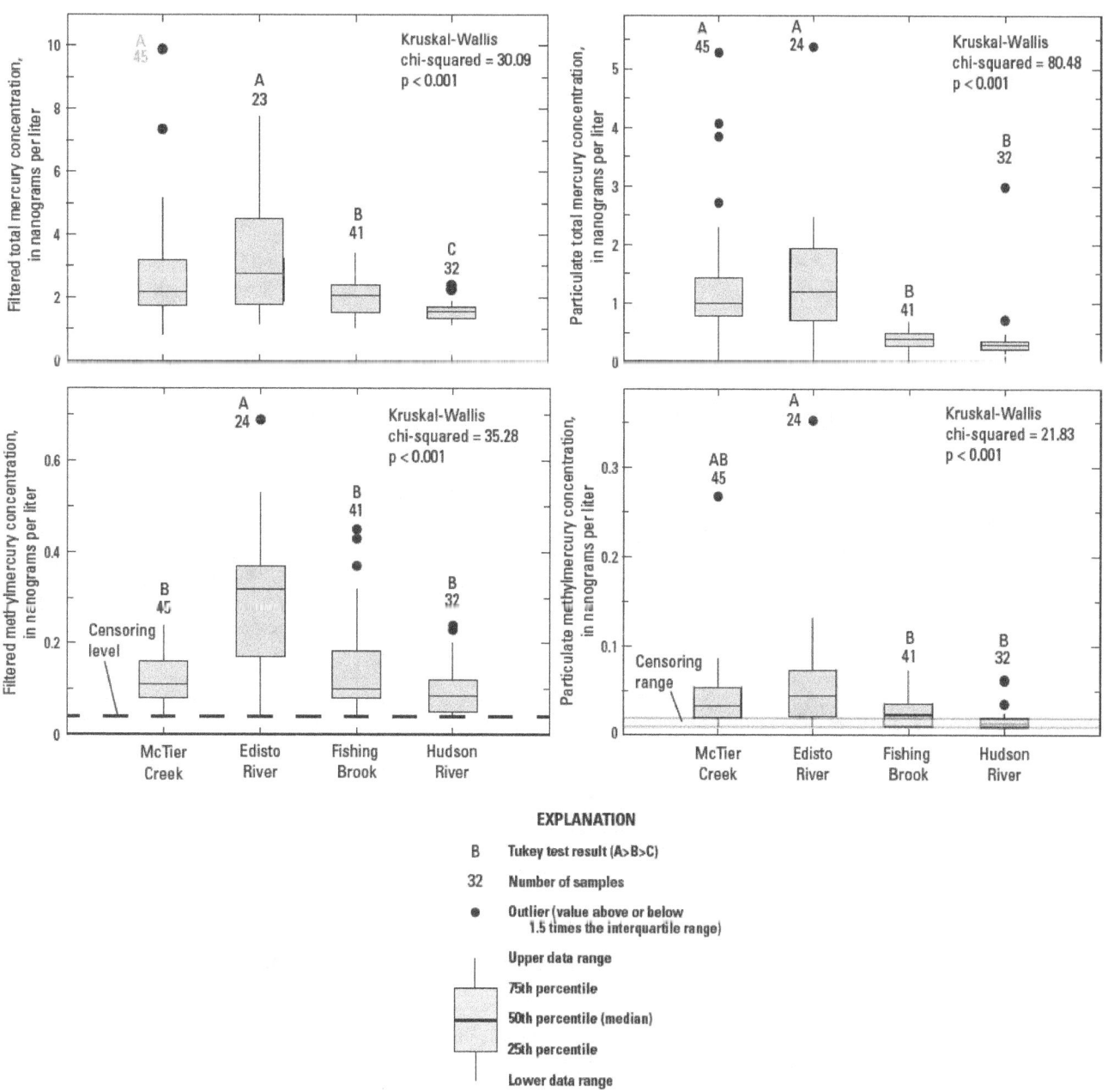

Figure 8. Distributions of filtered total mercury, particulate total mercury, filtered methylmercury, and particulate methylmercury for the McTier Creek near New Holland, S.C. (McTier Creek, station 02172305), Edisto River near Givhans, S.C. (Edisto River, station 02175000), Fishing Brook at County Line Flow near Newcomb, N.Y. (Fishing Brook, station 0131199050), and Hudson River near Newcomb, N.Y. (Hudson River, station 01312000) sites, 2007 to 2009.

Table 5. Spearman's correlation coefficients (rho) and probability values (p-values) between selected variables at McTier Creek near New Holland, S.C., Edisto River near Givhans, S.C., Fishing Brook at County Line Flow near Newcomb, N.Y., and Hudson River near Newcomb, N.Y., water years 2005 to 2009.

[<, less than; red highlighted text, statistically significant correlation]

Site		Spearman correlation with streamflow								
		Filtered total mercury	Filtered methyl-mercury	Particulate total mercury	Particulate methyl-mercury	Combined filtered and particulate methylmercury	Combined filtered and particulate total mercury	Dissolved organic carbon	Suspended sediment	Suspended sediment as percent finer than 63 microns
McTier Creek	rho	0.561	−0.493	0.371	−0.067	−0.328	0.528	0.565	0.148	−0.216
	p-value	<0.001	<0.001	0.012	0.662	0.028	0.000	<0.001	0.339	0.193
Edisto River	rho	0.729	0.506	0.196	0.276	0.489	0.631	0.733	0.051	−0.131
	p-value	<0.001	0.012	0.360	0.192	0.015	0.001	<0.001	0.671	0.403
Fishing Brook	rho	0.346	−0.551	−0.103	−0.532	−0.579	0.245	−0.342	−0.415	−0.150
	p-value	0.027	<0.001	0.520	<0.001	<0.001	0.123	0.025	0.007	0.428
Hudson River	rho	0.607	−0.568	0.831	0.261	−0.490	0.757	0.055	0.230	−0.657
	p-value	<0.001	<0.001	<0.001	0.150	0.004	<0.0001	0.753	0.191	0.020

streamflow (table 5; FTHg p-value = 0.027; FMeHg p-value < 0.001). Unlike at the South Carolina sites, DOC concentrations appeared to respond to streamflow in the same way as FMeHg at the Fishing Brook site, producing a significant negative correlation (table 5; p-value = 0.025). The Fishing Brook site was the only site that had PMeHg concentrations that were correlated strongly and negatively to streamflow (table 5; PMeHg p-value < 0.001). Typically, greater amounts of particulate material are delivered to the stream during storm runoff, thus producing an increased concentration with streamflow. Unlike at the South Carolina sites, DOC and SSC concentrations appeared to respond to streamflow in the same way as FMeHg at the Fishing Brook site, producing a significant negative correlation (table 5; DOC p-value = 0.025; SSC p-value = 0.007). Unlike the Edisto River site in South Carolina, the Hudson River site had FMeHg concentrations that responded in the same way to streamflow as its headwater catchment site, Fishing Brook. At the Hudson River site, FTHg concentrations were positively correlated to streamflow, and FMeHg concentrations were negatively correlated to streamflow (table 5; FTHg p-value < 0.001; FMeHg p-value < 0.001).

Spearman's rho correlation also was used to determine the relation among mercury species and selected water-quality constituents, including water temperature (related to seasonality and biological activity), DOC, ultraviolet absorbance at 254 nanometers (nm) (UVA254; estimate of the terrestrially derived humic fraction of DOC), particulate organic carbon (POC), nutrients, and seston chlorophyll *a* and pheophytin *a* (estimate of algae). At all sites, DOC and UVA254 were positively correlated to FTHg (table 6). Only Edisto River and Fishing Brook had significant correlations between FMeHg and DOC and UVA254. Particulate forms of Hg also were positively correlated with DOC and UVA254 at all sites except

the Hudson River site. At the Hudson River, Fishing Brook, and McTier Creek sites, POC had a stronger correlation to PTHg and PMeHg than DOC.

An interesting finding at the Fishing Brook site was the significant and positive correlation between PMeHg and PTHg to chlorophyll *a* concentrations (table 6). The Fishing Brook site is located at the outlet of a large run-of-river–type impoundment. This pattern of greater DOC, suspended material, and PMeHg concentrations during lower streamflows, therefore, could be influenced by algal-derived (autochthonous) particulates rather than terrestrially derived (allochthonous) particulates as a major source. Whereas the McTier Creek site had PMeHg concentrations positively correlated to algal-associated variables, the Fishing Brook site had the strongest positive correlation (p < 0.001) to water temperature, total phosphorus, and chlorophyll *a* concentrations (table 6), suggesting some influence by algal activity. At the Hudson River site, PTHg concentrations were correlated positively to streamflow (table 5; PTHg p-value < 0.001), but PMeHg, DOC, and SSC concentrations were not significantly correlated to streamflow.

Estimated Annual Mercury Loads and Yields

Annual loads for water years 2008 and 2009 were averaged and compared between headwater and large river sites in each basin (table 7; appendix 3-A). During the 2 water years, drought conditions in the Edisto River basin and above-average flow conditions in the upper Hudson River basin produced different streamflow regimes that influenced the load results; therefore, the 2-year mean annual load estimates were not compared between the Edisto River and upper Hudson River basins.

Table 6. Spearman's correlation coefficients (rho) and probability values (p-values) between mercury species concentrations at McTier Creek near New Holland, S.C., Edisto River near Givhans, S.C., Fishing Brook at County Line Flow near Newcomb, N.Y., and Hudson River near Newcomb, N.Y., and selected correlative variables, water years 2005 to 2009.

[ND, no data; <, less than; red highlighted text, statistically significant correlation; nm, nanometers]

Site		Water temperature	Dissolved organic carbon	Ultraviolet absorbance at 254 nm	Particulate organic carbon	Particulate nitrogen	Nitrate plus nitrite	Total nitrogen	Total phosphorus	Pheophytin a (Seston)	Chlorophyll a (Seston)
\multicolumn{12}{c}{Correlation with filtered total mercury (FTHg)}											
McTier	rho	−0.305	0.327	0.320	−0.068	−0.073	0.359	0.276	−0.079	−0.021	−0.122
Creek	p-value	0.050	0.034	0.041	0.690	0.670	0.018	0.074	0.610	0.900	0.460
Edisto	rho	−0.373	0.707	0.731	0.000	0.159	−0.609	0.301	−0.018	ND	ND
River	p-value	0.087	<0.001	0.000	1.000	0.680	0.003	0.180	0.940	ND	ND
Fishing	rho	0.315	0.553	0.657	0.200	0.155	−0.133	−0.056	0.256	0.034	0.089
Brook	p-value	0.045	<0.001	<0.001	0.210	0.330	0.410	0.730	0.110	0.860	0.630
Hudson	rho	0.126	0.426	0.666	0.501	0.437	−0.014	0.127	0.185	0.383	0.237
River	p-value	0.490	0.015	<0.001	0.004	0.012	0.940	0.490	0.310	0.250	0.480
\multicolumn{12}{c}{Correlation with filtered methylmercury (FMeHg)}											
McTier	rho	0.535	0.055	0.162	0.223	0.017	−0.295	0.117	0.266	0.361	0.181
Creek	p-value	<0.001	0.722	0.300	0.170	0.920	0.049	0.450	0.077	0.021	0.260
Edisto	rho	−0.040	0.815	0.832	0.417	0.720	−0.308	0.579	0.238	ND	ND
River	p-value	0.860	<0.001	<0.001	0.260	0.029	0.160	0.005	0.290	ND	ND
Fishing	rho	0.697	0.673	0.672	0.610	0.701	0.214	−0.145	0.687	0.682	0.581
Brook	p-value	<0.001	<0.001	<0.001	<0.001	<0.001	0.190	0.370	<0.001	<0.001	<0.001
Hudson	rho	0.699	0.296	0.151	0.214	0.108	−0.754	−0.722	0.242	0.443	0.270
River	p-value	<0.001	0.100	0.410	0.240	0.560	<0.001	<0.001	0.180	0.170	0.420
\multicolumn{12}{c}{Correlation with particulate total mercury (PTHg)}											
McTier	rho	0.236	0.495	0.500	0.485	0.505	0.181	0.658	0.405	0.401	0.228
Creek	p-value	0.123	<0.001	<0.001	0.002	<0.001	0.230	<0.001	0.006	0.009	0.150
Edisto	rho	0.256	0.422	0.459	0.267	0.519	0.067	0.761	0.663	ND	ND
River	p-value	0.240	0.045	0.028	0.490	0.150	0.770	<0.001	<0.001	ND	ND
Fishing	rho	0.418	0.445	0.455	0.690	0.614	−0.560	−0.088	0.687	0.513	0.617
Brook	p-value	0.007	0.004	0.003	<0.001	<0.001	<0.001	0.590	<0.001	0.003	<0.001
Hudson	rho	−0.088	0.070	0.247	0.545	0.571	0.300	0.379	0.195	0.273	0.232
River	p-value	0.630	0.710	0.170	0.001	<0.001	0.096	0.033	0.280	0.420	0.490
\multicolumn{12}{c}{Correlation with particulate methylmercury (PMeHg)}											
McTier	rho	0.574	0.408	0.456	0.528	0.312	−0.227	0.528	0.593	0.529	0.378
Creek	p-value	<0.001	0.006	0.002	<0.001	0.050	0.130	<0.001	<0.001	<0.001	0.015
Edisto	rho	0.363	0.530	0.517	0.468	0.710	0.041	0.873	0.752	ND	ND
River	p-value	0.089	0.009	0.012	0.200	0.032	0.860	<0.001	<0.001	ND	ND
Fishing	rho	0.826	0.696	0.693	0.894	0.866	−0.814	−0.024	0.902	0.828	0.822
Brook	p-value	<0.001	<0.001	<0.001	<0.001	<0.001	<0.001	0.880	<0.001	<0.001	<0.001
Hudson	rho	0.496	0.333	0.240	0.793	0.774	−0.352	−0.231	0.212	0.538	0.362
River	p-value	0.004	0.063	0.190	<0.001	<0.001	0.048	0.200	0.243	0.088	0.274

The watershed of the McTier Creek headwater stream site composed 1.1 percent of the basin for the Edisto River site. The 2-year mean annual FTHg load of 0.0663 kilogram per year (kg/yr) at the McTier Creek site represented about 1.4 percent (similar to the basin size percentage) of the mean annual FTHg load of 4.83 kg/yr at the Edisto River site (table 7). Similarly, the mean annual PTHg load of 0.0273 kg/yr at the McTier Creek site represented about 1.4 percent of the mean annual PTHg load of 1.91 kg/yr at the Edisto River site. These load comparisons suggest that THg contribution in the Edisto River basin remained relatively constant between headwater stream and larger river basin, with no change in THg contribution with increased basin scale. However, the mean annual FMeHg load of 0.0020 kg/yr at the McTier Creek site composed a much lower percentage (about 0.5 percent) (lower than the basin size percentage) of the mean annual FMeHg of 0.449 kg/yr at the Edisto River site. The mean annual PMeHg load of 0.00070 kg/yr at the McTier Creek site represented 0.2 percent of the mean annual PMeHg load of 0.0782 kg/yr at the Edisto River site. Based on these load comparisons, contribution of MeHg was greater at the larger river site than at the headwater stream, suggesting that MeHg contribution increased with basin scale in the Edisto River basin.

The watershed of the Fishing Brook headwater stream site composed about 13 percent of the basin for the Hudson River site. As was determined for the Edisto River basin sites, the 2-year mean annual FTHg load of 0.109 kg/yr at the Fishing Brook site represented 13 percent (similar to the basin size percentage) of the mean annual FTHg load of 0.827 kg/yr at the Hudson River site (table 7). The mean annual PTHg load of 0.020 kg/yr at the Fishing Brook site represented only 9 percent of the mean annual PTHg load of 0.221 kg/yr at the Hudson River site. Unlike the Edisto River basin sites, mean annual FMeHg load of 0.00619 kg/yr at the Fishing Brook site composed more than 18 percent (greater than the basin size percentage) of the mean annual FMeHg load of 0.0336 kg/yr at the Hudson River site. The mean annual PMeHg load of 0.00104 kg/yr at the Fishing Brook site represented 11 percent of the mean annual PMeHg load of 0.00910 kg/yr at the Hudson River site. Based on these load comparisons, contribution of FMeHg was greater and contribution of particulate Hg species was less at the smaller scale headwater site than at the large downstream site.

Table 7. Estimated mean annual loads and yields of fluvial total mercury and methylmercury for water years 2008 and 2009 at McTier Creek near New Holland, S.C. (McTier Creek, 02172305), Edisto River near Givhans, S.C. (Edisto River, 02175000), Fishing Brook at County Line Flow near Newcomb, N.Y. (Fishing Brook, 0131199050), and Hudson River near Newcomb, N.Y. (Hudson River, 01312000).

[THg, total mercury; MeHg, methylmercury; kg/yr, kilograms per year; (µg/m²)/yr, micrograms per square meter per year. Total mercury includes methylated and inorganic forms of mercury. A water year extends from October of one calendar year to September of the following calendar year]

Site	State	2-Year mean annual loads (kg/yr)			2-Year mean annual yields [(µg/m²)/yr]		
		Filtered THg	Particulate THg	THg	Filtered THg	Particulate THg	THg
McTier Creek	S.C.	0.066	0.027	0.094	0.834	0.343	1.18
Edisto River	S.C.	4.83	1.91	6.74	0.683	0.270	0.953
Fishing Brook	N.Y.	0.109	0.020	0.129	1.67	0.304	1.98
Hudson River	N.Y.	0.827	0.221	1.05	1.66	0.444	2.11

Site	State	2-Year mean annual loads (kg/yr)			2-Year mean annual yields [(µg/m²)/yr]		
		Filtered MeHg	Particulate MeHg	MeHg	Filtered MeHg	Particulate MeHg	MeHg
McTier Creek	S.C.	0.002	0.001	0.003	0.025	0.009	0.034
Edisto River	S.C.	0.449	0.078	0.527	0.064	0.011	0.075
Fishing Brook	N.Y.	0.006	0.001	0.007	0.095	0.016	0.111
Hudson River	N.Y.	0.034	0.009	0.043	0.068	0.018	0.086

Annual yields, which are the annual loads divided by drainage basin size, computed for water years 2008 and 2009 were used to provide a more direct evaluation than annual loads for the comparison of contributions between headwater and large river sites in each basin. Because of differences in climatic conditions between the two basins during these 2 water years, yields were not compared between basins. The estimated 2-year mean annual FMeHg yield (water years 2008 and 2009) at the McTier Creek site of 0.0252 microgram per square meter per year ($\mu g/m^2$)/yr was almost 3 times lower than the FMeHg yield at the Edisto River site of 0.0635 ($\mu g/m^2$)/yr, indicative of increased contributions with increasing scale from headwater stream catchment to larger river basin (fig. 9; table 7). However, the year-to-year variability in the Edisto River FMeHg yield also was greater compared to that of the McTier Creek site. Two-year mean annual yields of FTHg were 0.834 and 0.683 ($\mu g/m^2$)/yr at the McTier and Edisto sites, respectively, indicative of negligible change in contribution from headwater stream to larger river scale. Yields of particulate forms of mercury were relatively consistent between headwater stream and larger river basin scales. Two-year mean annual yields of PMeHg were 0.0088 and 0.0111 ($\mu g/m^2$)/yr at the McTier Creek and Edisto River sites, respectively. Annual PTHg yields of 0.343 ($\mu g/m^2$)/yr at the McTier Creek site and 0.270 ($\mu g/m^2$)/yr at the Edisto River site were similar between sites.

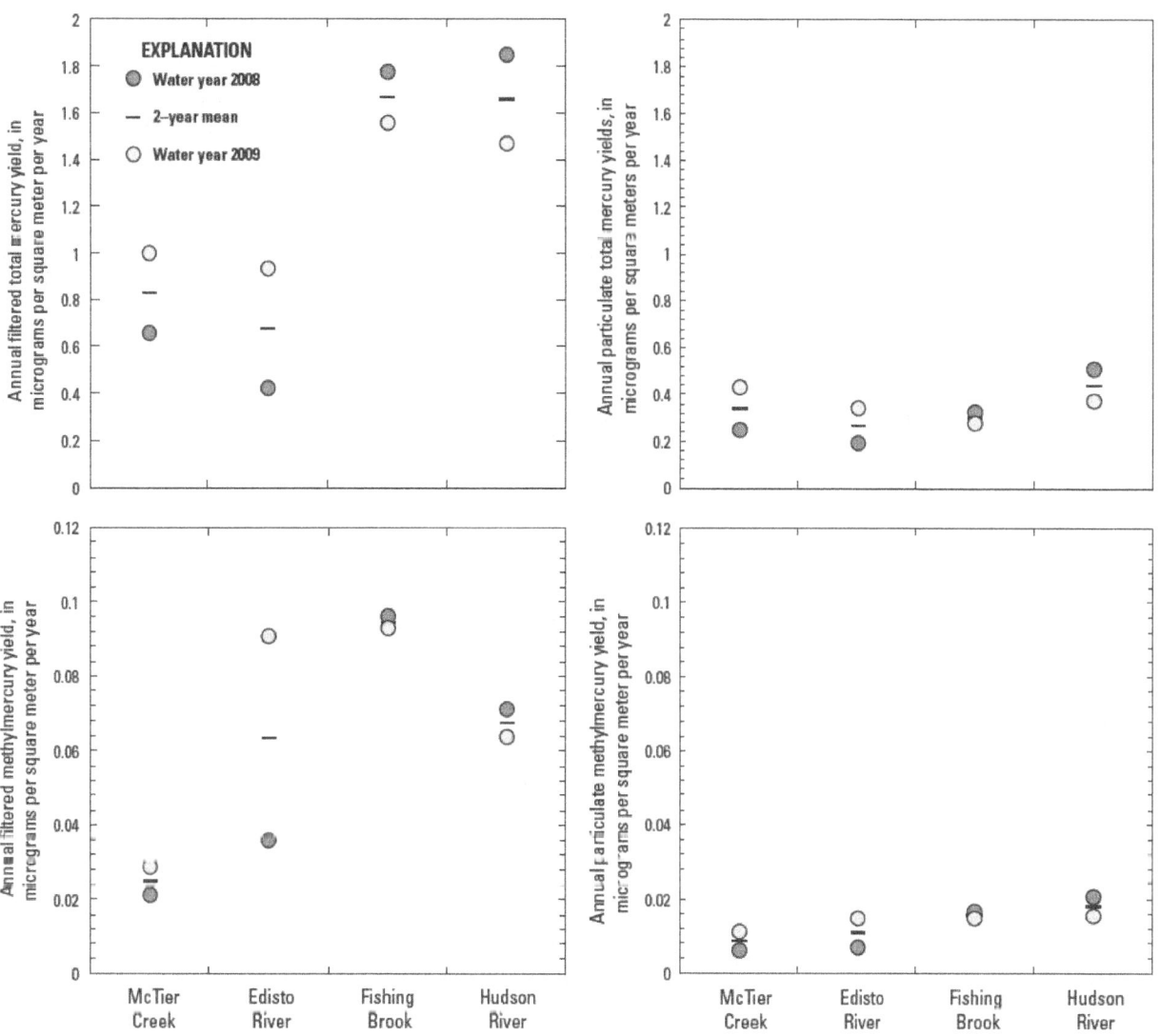

Figure 9.   Annual filtered total mercury, particulate total mercury, filtered methylmercury, and particulate methylmercury yields for water years 2008 and 2009 for McTier Creek near New Holland, S.C. (McTier Creek, station 02172305), Edisto River near Givhans, S.C. (Edisto River, station 02175000), Fishing Brook at County Line Flow near Newcomb, N.Y. (Fishing Brook, station 0131199050), and Hudson River near Newcomb, N.Y. (Hudson River, station 01312000) sites.

In contrast to the South Carolina sites, the FMeHg yield of 0.0948 (μg/m^2)/yr at the Fishing Brook site in New York was higher than the FMeHg yield of 0.0676 (μg/m^2)/yr at the Hudson River site, indicating decreased contributions with increasing scale from headwater stream catchment to larger river basin (fig. 9; table 7). Similar to the South Carolina sites, the 2-year mean annual yields of FTHg at the Fishing Brook and Hudson River sites indicated no change in contribution from headwater stream to larger river scale and were 1.67 and 1.66 (μg/m^2)/yr, respectively. Two-year mean annual yields of PMeHg were 0.0159 and 0.0183 (μg/m^2)/yr at the Fishing Brook and Hudson River sites, respectively, but the 2-year mean annual yield of PTHg at the Fishing Brook site of 0.304 (μg/m^2)/yr was 32 percent lower than the mean annual PTHg yield of 0.444 (μg/m^2)/yr at the Hudson River site.

Mean annual loads and yields that were computed for water years 2005 to 2009 represented a larger range of flow conditions in each basin and allowed comparison of mean annual yields among sites in the two basins over more climatically similar conditions (appendix 3-A). Streamflow was estimated using the MOVE.1 technique for water years 2005 to 2007 at the McTier Creek and Fishing Brook sites and used as input into the load model. Calibration datasets collected at the Edisto River and Hudson River sites extended throughout this period. Calibration datasets collected at the McTier Creek and Fishing Brook sites were limited only to the 2007 to 2009 study period, which increased the level of uncertainty in the model estimates. This increased uncertainty was most pronounced for the McTier Creek site because of the low-flow conditions under which the calibration dataset was collected (appendix 3-A).

During the 5-year period, relatively similar mean annual FTHg yields were observed at the paired (headwater stream/large river) basin sites in New York and South Carolina; however, differences were observed between basins. Mean annual FTHg yields of 1.01 and 0.754 (μg/m^2)/yr at the McTier Creek and Edisto River paired basin sites, respectively, were lower than the mean annual FTHg yields of 1.64 and 1.60 (μg/m^2)/yr at the Fishing Brook and Hudson River paired basin sites, respectively (fig. 10; table 8). The McTier Creek site also had a mean annual FMeHg yield of 0.0297 (μg/m^2)/yr that was 2 to 3 times lower than the mean annual FMeHg yields of 0.0732 (μg/m^2)/yr at the Edisto River site, 0.0919 (μg/m^2)/yr at the Fishing Brook site, and 0.0632 (μg/m^2)/yr at the Hudson River site. Particulate species of mercury had lower yields than filtered species at all sites, and yields were relatively similar among the four sites. Mean annual PTHg yields ranged from 0.288 (μg/m^2)/yr (Edisto River) to 0.454 (μg/m^2)/yr (McTier Creek) among sites (table 8, fig. 10). Mean annual PMeHg yields ranged from 0.0123 (μg/m^2)/yr (McTier Creek, Edisto River) to 0.0170 (μg/m^2)/yr (Hudson River).

During water years 2005 to 2009, mean annual DOC yields followed a similar spatial trend among sites as those of mean annual FTHg. Relatively similar mean annual DOC yields were observed at the paired (headwater stream/large river) basin sites in New York and South Carolina; however,

differences were observed between basins. Mean annual DOC yields of 24.4 and 16.7 kilograms per hectare per year [(kg/ha)/yr] at the McTier Creek and Edisto River paired basin sites, respectively, were lower than the mean annual DOC yields of 54.4 and 52.9 (kg/ha)/yr at the Fishing Brook and Hudson River paired basin sites, respectively (fig. 11; table 9). Mean annual yields of dissolved chloride and sulfate also were lower at the McTier Creek and Edisto River paired basin sites when compared to those of the Fishing Brook and Hudson River paired basin sites. In South Carolina, mean annual dissolved chloride yields were 9.60 (kg/ha)/yr at the headwater McTier Creek site and 14.6 (kg/ha)/yr at the larger Edisto River site, indicating a slight increase with basin scale. Conversely, in New York, mean annual dissolved chloride yields were 39.2 (kg/ha)/yr at the headwater Fishing Brook site and 24.2 (kg/ha)/yr at the larger Hudson River site, indicating a slight decrease with basin scale. However, basin-scale changes in mean annual dissolved sulfate yields were more consistent between the two paired basins and indicated increasing yields with increasing basin scales. In South Carolina, the mean annual dissolved sulfate yield of 5.30 (kg/ha)/yr at the McTier Creek site was half the mean annual sulfate yield at the Edisto River site [10.5 (kg/ha)/yr]. In New York, the Fishing Brook site had a mean annual dissolved sulfate yield of 29.1 (kg/ha)/yr, which was about 60 percent lower than the mean annual sulfate yield of 49.1 (kg/ha)/yr at the Hudson River site.

Mean annual SSC yields did not exhibit a consistent pattern between the New York and South Carolina basins or between paired (headwater/large river) sites. The mean annual SSC yield of 2.5 (kg/ha)/yr at the New York headwater site, Fishing Brook, was the lowest by an order of magnitude, compared with the other three study sites, and the South Carolina headwater site, McTier Creek, had the highest mean annual SSC yield for the four study sites [31.0 (kg/ha)/yr]. The Edisto River and Hudson River sites appeared to have relatively similar mean annual SSC yields of 14.7 and 21.0 (kg/ha)/yr, respectively, based on overlapping 95 percent confidence intervals.

Comparison of Stream Yields to Atmospheric Deposition

Total (filtered plus particulate) Hg, dissolved chloride, and dissolved sulfate yields in the streams at the study sites were compared to Mercury Deposition Network (MDN; mercury) and National Trends Network (NTN; chloride and sulfate) wet atmospheric deposition and litterfall (THg only) to determine how much of the atmospheric deposition reaches the stream. MeHg was not measured in this study; however, MeHg has been shown to account for only a small percentage (0.6 to 1.5 percent) of the THg measured in composite litterfall samples in the eastern United States (Risch and others, 2012). Annual THg deposition from litterfall also was computed for water year 2007 in both basins to provide an estimate of the dry deposition of THg. Litterfall is thought to capture gaseous

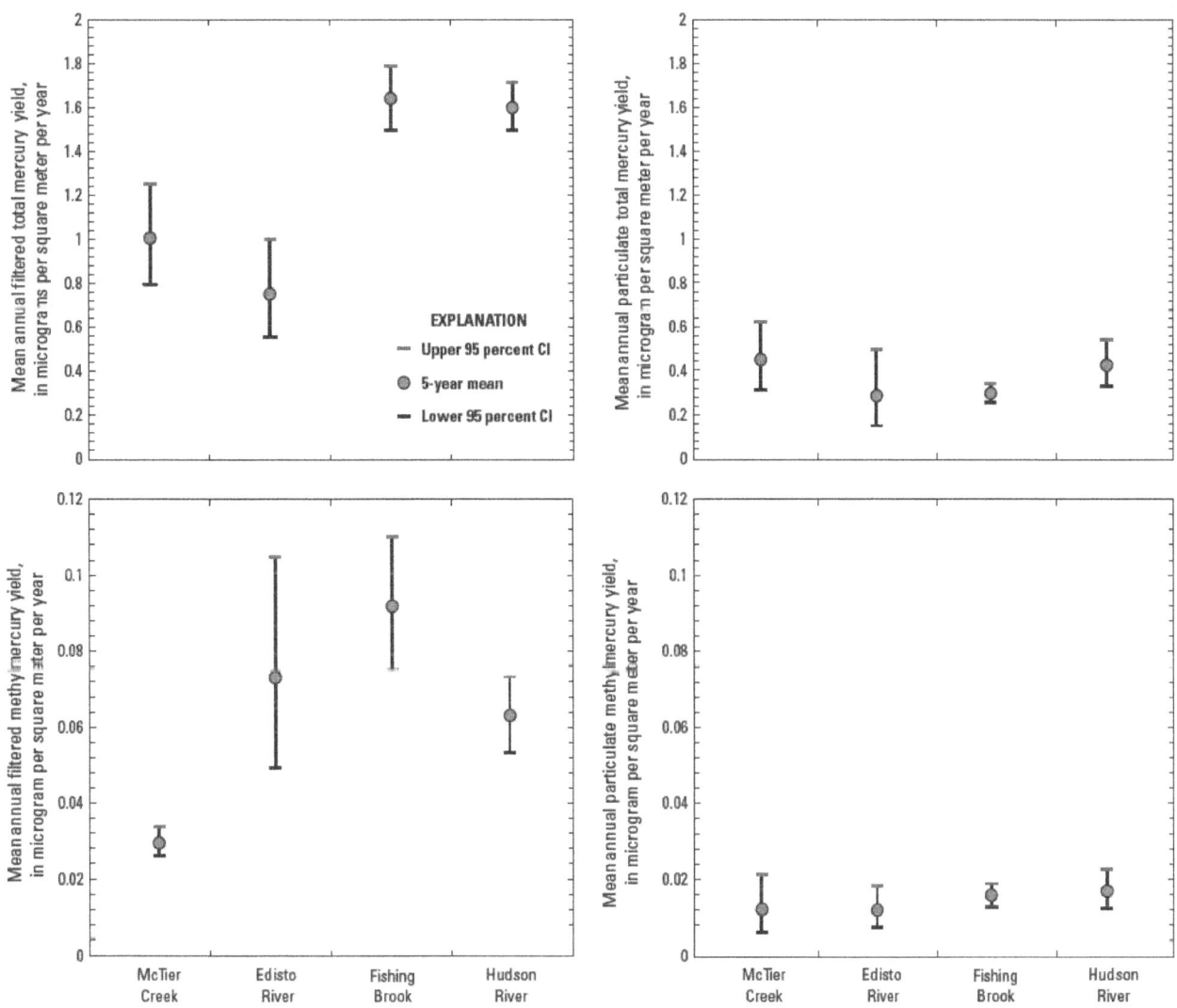

Figure 10. Mean annual and 95 percent confidence levels (CI) of filtered total mercury, particulate total mercury, filtered methylmercury, and particulate methylmercury yields for McTier Creek near New Holland, S.C. (McTier Creek, station 02172305), Edisto River near Givhans, S.C. (Edisto River, station 02175000), Fishing Brook at County Line Flow near Newcomb, N.Y. (Fishing Brook, station 0131199050), and Hudson River near Newcomb, N.Y. (Hudson River, station 01312000) sites, water years 2005 to 2009.

Hg that is taken up through the leaf stomata during growing season and the reactive or particulate Hg that sticks to leaf surfaces (St. Louis and others, 2001; Risch and others, 2012). For the eastern United States, mean annual litterfall Hg deposition has been shown to be significantly higher than mean annual wet deposition of mercury (Risch and others, 2012); therefore, litterfall contributions should be included in the total THg deposition estimate.

In the McTier Creek headwater stream basin, the PRISM-estimated annual wet deposition of THg ranged from 7.60 to 12.2 (µg/m²)/yr and averaged 9.91 (µg/m²)/yr (fig. 12; table 10). Litterfall accounted for an estimated 12.8 (µg/m²)/yr of THg deposition in this basin in 2007, which produced a

total (wet plus litterfall) THg deposition of 20.4 (µg/m²)/yr (fig. 13; table 10). Based on these estimates, wet deposition of THg represented only 37.3 percent of the total THg deposition in the McTier Creek basin in 2007 (fig. 13; table 10). Annual THg wet deposition and litterfall deposition at the Cape Romain National Wildlife Refuge MDN site SC05, located outside of the study area in Charleston County, S.C., were estimated for 2008 and 2009 (Risch and others, 2012) and were used for comparison to the estimates from this study (table 10). In 2008 and 2009, annual THg wet depositions at the MDN SC05 site were lower than the PRISM-estimated THg wet depositions at the McTier Creek site, especially in 2009 [12.2 (µg/m²)/yr, McTier Creek; 7.4 (µg/m²)/yr,

Table 8. Estimated mean annual loads and yields of fluvial total mercury (THg) and methylmercury (MeHg) for water years 2005 to 2009, with lower and upper bounds of the 95 percent confidence interval (CI) for the model at McTier Creek near New Holland, S.C. (McTier Creek, 02172305), Edisto River near Givhans, S.C. (Edisto River, 02175000), Fishing Brook at County Line Flow near Newcomb, N.Y. (Fishing Brook, 0131199050), and Hudson River near Newcomb, N.Y. (Hudson River, 01312000).

[For the McTier Creek and Fishing Brook sites, discharge records were extended to water years 2005, 2006, and 2007 using MOVE.1 estimation techniques; therefore, the 2005, 2006, and 2007 annual loads at these two sites were estimated using the extended discharge record. Gaged discharge data were used for load computations for all annual loads at the Edisto River and Hudson River sites and for 2008 and 2009 annual loads for the McTier Creek and Fishing Brook sites. Total mercury includes methylated and inorganic forms of mercury. FTHg, filtered total mercury; PTHg, particulate total mercury; FMeHg, filtered methylmercury; PMeHg, particulate methylmercury; A water year extends from October of one calendar year to September of the following calendar year; %, percent; kg/yr, kilograms per year; (μg/m²)/yr, micrograms per square meter per year]

Total mercury (THg) loads

Site	State	Mean annual FTHg loads			Mean annual PTHg loads			Mean annual THg loads		
		Load (kg/yr)	Lower 95% CI (kg/yr)	Upper 95% CI (kg/yr)	Load (kg/yr)	Lower 95% CI (kg/yr)	Upper 95% CI (kg/yr)	Load (kg/yr)	Lower 95% CI (kg/yr)	Upper 95% CI (kg/yr)
McTier Creek	S.C.	0.080	0.064	0.099	0.036	0.025	0.050	0.116	0.089	0.149
Edisto River	S.C.	5.33	3.94	7.07	2.04	1.08	3.53	7.37	5.02	10.6
Fishing Brook	N.Y.	0.107	0.098	0.117	0.020	0.017	0.023	0.127	0.115	0.139
Hudson River	N.Y.	0.796	0.745	0.850	0.213	0.165	0.271	1.01	0.910	1.121

Methylmercury (MeHg) loads

Site	State	Mean annual FMeHg loads			Mean annual PMeHg loads			Mean annual MeHg loads		
		Load (kg/yr)	Lower 95% CI (kg/yr)	Upper 95% CI (kg/yr)	Load (kg/yr)	Lower 95% CI (kg/yr)	Upper 95% CI (kg/yr)	Load (kg/yr)	Lower 95% CI (kg/yr)	Upper 95% CI (kg/yr)
McTier Creek	S.C.	0.002	0.002	0.003	0.001	0.001	0.002	0.003	0.003	0.004
Edisto River	S.C.	0.518	0.351	0.741	0.086	0.053	0.132	0.604	0.404	0.873
Fishing Brook	N.Y.	0.006	0.005	0.007	0.001	0.001	0.001	0.007	0.006	0.009
Hudson River	N.Y.	0.031	0.027	0.037	0.009	0.006	0.011	0.040	0.028	0.048

Total mercury (THg) yields, by water year

Site	State	Mean annual FTHg yields			Mean annual PTHg yields			Mean annual THg yields		
		Yield [(μg/m²)/yr]	Lower 95% CI [(μg/m²)/yr]	Upper 95% CI [(μg/m²)/yr]	Yield [(μg/m²)/yr]	Lower 95% CI [(μg/m²)/yr]	Upper 95% CI [(μg/m²)/yr]	Yield [(μg/m²)/yr]	Lower 95% CI [(μg/m²)/yr]	Upper 95% CI [(μg/m²)/yr]
McTier Creek	S.C.	1.01	0.799	1.25	0.454	0.317	0.627	1.46	1.12	1.88
Edisto River	S.C.	0.754	0.557	1.00	0.288	0.153	0.499	1.04	0.710	1.50
Fishing Brook	N.Y.	1.64	1.50	1.79	0.301	0.259	0.347	1.94	1.76	2.14
Hudson River	N.Y.	1.60	1.50	1.71	0.428	0.332	0.545	2.03	1.83	2.25

Methylmercury (MHg) yields, by water year

Site	State	Mean annual FMeHg yields			Mean annual PMeHg yields			Mean annual MeHg yields		
		Yield [(μg/m²)/yr]	Lower 95% CI [(μg/m²)/yr]	Upper 95% CI [(μg/m²)/yr]	Yield [(μg/m²)/yr]	Lower 95% CI [(μg/m²)/yr]	Upper 95% CI [(μg/m²)/yr]	Yield [(μg/m²)/yr]	Lower 95% CI [(μg/m²)/yr]	Upper 95% CI [(μg/m²)/yr]
McTier Creek	S.C.	0.030	0.026	0.033	0.012	0.006	0.021	0.042	0.033	0.054
Edisto River	S.C.	0.073	0.050	0.105	0.012	0.008	0.019	0.085	0.057	0.123
Fishing Brook	N.Y.	0.092	0.075	0.110	0.016	0.013	0.019	0.107	0.088	0.130
Hudson River	N.Y.	0.063	0.054	0.073	0.017	0.013	0.023	0.080	0.056	0.096

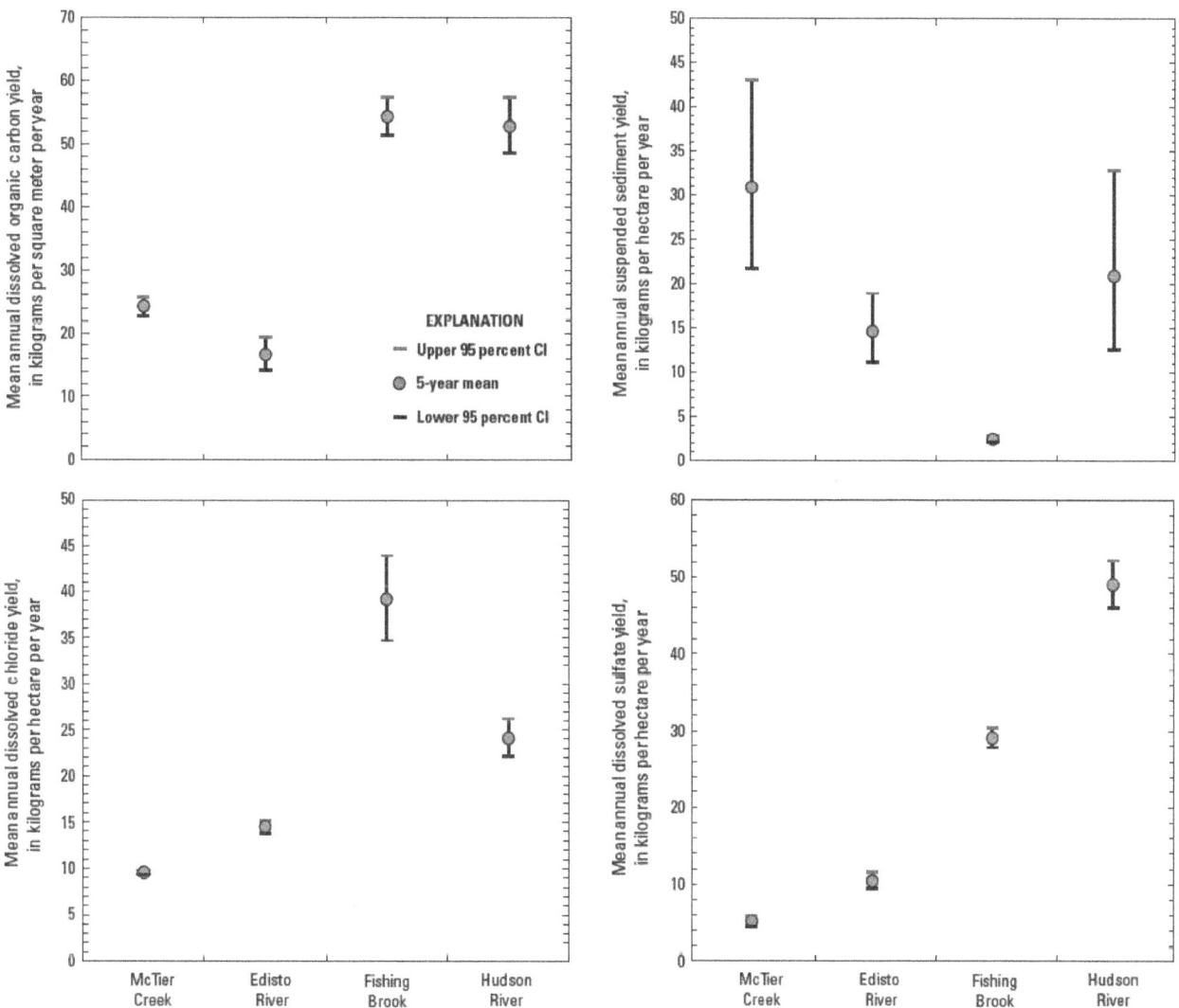

Figure 11. Mean annual and 95 percent confidence levels (CI) of dissolved organic carbon, suspended sediment, dissolved chloride, and dissolved sulfate yields for the McTier Creek near New Holland, S.C. (McTier Creek, station 02172305), Edisto River near Givhans, S.C. (Edisto River, station 02175000), Fishing Brook at County Line Flow near Newcomb, N.Y. (Fishing Brook, station 0131199050), and Hudson River near Newcomb, N.Y. (Hudson River, station 01312000) sites, water years 2005 to 2009.

MDN SC05; table 10]. Mean wet deposition of THg of 8.2 $(\mu g/m^2)$/yr was almost half (47 percent) of the mean total deposition (litterfall plus wet deposition) of THg of 17.5 $(\mu g/m^2)$/yr at the MDN SC05 site (table 10; Risch and others, 2012). Although wet deposition estimates were slightly lower than those for the McTier Creek basin, the MDN SC05 estimates also indicated that dry deposition of THg related to litterfall was a significant source of atmospherically derived THg to the MDN SC05 basin.

The Fishing Brook headwater stream basin had lower PRISM-estimated annual wet deposition of THg than the McTier Creek site. The annual wet deposition of THg ranged from 6.24 to 7.17 $(\mu g/m^2)$/yr and averaged 6.30 $(\mu g/m^2)$/yr for water years 2005 to 2009 in this basin (fig. 12; table 10). Litterfall accounted for an estimated 9.00 $(\mu g/m^2)$/yr of THg in this basin in 2007, which produced a total THg deposition

of 15.4 $(\mu g/m^2)$/yr (fig. 13; table 10). Based on these estimates, wet deposition of THg represented only 41.6 percent of the total THg deposition in the Fishing Brook basin in 2007 (fig. 13; table 10).

Annual THg wet deposition and litterfall deposition at the Biscuit Brook MDN site NY68, located outside the study area in Ulster County, N.Y., were estimated for 2008 and 2009 and were consistent with the PRISM-derived estimates from this study (fig. 13; table 10; Risch and others, 2012). In 2008 and 2009, annual THg wet depositions at the MDN NY68 site were higher than the PRISM-estimated THg wet depositions at the Fishing Brook site, especially in 2008 [7.17 $(\mu g/m^2)$/yr, Fishing Brook; 10.6 $(\mu g/m^2)$/yr, MDN NY68; fig. 13; table 10]. Mean of the 2008 and 2009 wet deposition of THg of 9.6 $(\mu g/m^2)$/yr was 38.6 percent of the mean total deposition of THg of 24.9 $(\mu g/m^2)$/yr at the MDN NY68 site

Table 9. Estimated mean annual loads and yields of fluvial dissolved organic carbon (DOC), suspended sediment (SSC), dissolved chloride, and dissolved sulfate for water years 2005 to 2009, with lower and upper bounds of the 95 percent confidence interval (CI) for the model at McTier Creek near New Holland, S.C (McTier Creek, 02172305), Edisto River near Givhans, S.C. (Edisto River, 02175000), Fishing Brook at County Line Flow near Newcomb, N.Y. (Fishing Brook, 0131199050), and Hudson River near Newcomb, N.Y. (Hudson River, 01312000).

[For the McTier Creek and Fishing Brook sites, discharge records were extended to water years 2005, 2006, and 2007 using MOVE.1 estimation techniques; therefore, the 2005, 2006, and 2007 annual loads at these two sites were estimated using the extended discharge record. Gaged discharge data were used for load computations for all annual loads at Edisto River and Hudson River sites and for 2008 and 2009 annual loads for the McTier Creek and Fishing Brook sites. A water year extends from October of one calendar year to September of the following calendar year. kg/yr, kilograms per year; (kg/ha)/yr, kilograms per hectare per year]

Site	DOC load, in kg/yr			DOC yield, in (kg/ha)/yr		
	Load	Lower 95% CI	Upper 95% CI	Yield	Lower 95% CI	Upper 95% CI
McTier Creek	194,043	182,181	206,481	24.4	22.9	26.0
Edisto River	11,791,468	10,093,457	13,693,172	16.7	14.3	19.4
Fishing Brook	355,137	335,976	375,097	54.4	51.5	57.5
Hudson River	2,630,006	2,419,363	2,854,013	52.9	48.6	57.4

Site	SSC load, in kg/yr			SSC Yield, in (kg/ha)/yr		
	Load	Lower 95% CI	Upper 95% CI	Yield	Lower 95% CI	Upper 95% CI
McTier Creek	246,474	173,398	341,506	31.0	21.8	43.0
Edisto River	10,362,027	7,861,004	13,418,481	14.7	11.1	19.0
Fishing Brook	16,279	13,795	19,082	2.50	2.10	2.90
Hudson River	1,042,183	628,819	1,635,290	21.0	12.6	32.9

Site	Dissolved chloride load, in kg/yr			Dissolved chloride yield, in (kg/ha)/yr		
	Load	Lower 95% CI	Upper 95% CI	Yield	Lower 95% CI	Upper 95% CI
McTier Creek	76,390	74,710	78,098	9.60	9.40	9.80
Edisto River	10,307,889	9,826,971	10,806,010	14.6	13.9	15.3
Fishing Brook	255,503	226,998	286,577	39.2	34.8	44.0
Hudson River	1,202,455	1,104,012	1,307,277	24.2	22.2	26.3

Site	Dissolved sulfate load, in kg/yr			Dissolved sulfate yield, in (kg/ha)/yr		
	Load	Lower 95% CI	Upper 95% CI	Yield	Lower 95% CI	Upper 95% CI
McTier Creek	41,790	36,735	47,349	5.30	4.60	5.90
Edisto River	7,434,690	6,711,172	8,214,405	10.5	9.5	11.6
Fishing Brook	190,066	181,853	198,550	29.1	27.9	30.4
Hudson River	2,441,347	2,290,013	2,599,968	49.1	46.0	52.3

(table 10; Risch and others, 2012). Although wet deposition estimates were slightly higher than those for the Fishing Brook basin, the MDN NY68 estimates were similar to those for the Fishing Brook basin and also indicated that dry deposition of THg related to litterfall was a significant source of atmospherically derived THg.

At the McTier Creek site, annual THg (filtered plus particulate) stream yields represented 9 to 19 percent of the estimated annual THg being delivered to the basin by wet deposition (fig. 12; table 10). At the Fishing Brook site, annual THg stream yields represented 23 to 33 percent of the estimated THg being delivered to the basin annually by wet deposition (fig. 12; table 10). In 2007, only 7 percent of the total (litterfall deposition plus wet deposition) THg deposition in the McTier Creek basin and 13 percent of the total THg deposition in the Fishing Brook basin reached the stream site, indicating storage of the atmospherically derived THg within the basin (table 10).

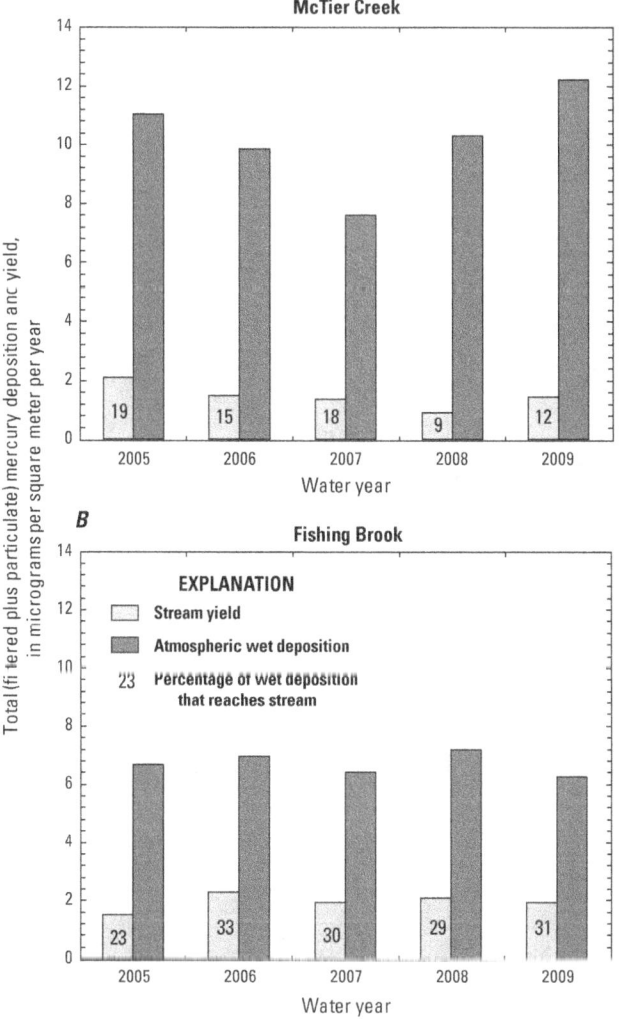

Figure 12. Estimated annual total (filtered plus particulate) mercury stream yields and annual total mercury wet deposition from litterfall and precipitation in the (*A*) McTier Creek near New Holland, S.C. (02172305) and (*B*) Fishing Brook at County Line Flow near Newcomb, N.Y. (0131199050) basins, water years 2005 to 2009.

Chloride is highly mobile and not involved in common aqueous geochemical reactions; therefore, chloride is considered to be a conservative element that tends to behave like water molecules during fluvial transport (Herczeg and Edmunds, 2000). Atmospheric wet deposition of chloride from precipitation to the McTier Creek and Fishing Brook basins was compared to annual chloride stream yields for water years 2005 to 2009 at these sites to evaluate to what degree atmospheric wet deposition contributed to stream yields (tables 9, 11; fig. 14). Annual stream yields of chloride at both sites were much higher than the chloride contributed to the basin

by annual wet deposition, suggesting that most of the chloride in the streams was not derived from atmospheric wet deposition (fig. 14; table 11). Annual wet deposition of chloride at the NADP SC06 site near the McTier Creek site ranged from 2.48 to 4.37 (kg/ha)/yr and had a mean of 3.71 (kg/ha)/yr (fig. 14; table 11). Annual stream yields of chloride at the McTier Creek site were about three times that of wet deposition and ranged from 6.98 to 12.4 (kg/ha)/yr, with a mean of 9.61 (kg/ha)/yr (fig. 14; table 11). Based on the mean annual estimates, only 39 percent of the mean annual stream yield of chloride can be attributed to atmospheric wet deposition of chloride in the McTier Creek basin (table 11). Annual wet deposition of chloride at the NADP NY20 site near the Fishing Brook site was typically lower than at the South Carolina NADP site, ranging from 0.56 to 1.2 (kg/ha)/yr and averaging 0.75 (kg/ha)/yr (fig. 14; table 11). Annual stream yields of chloride at the Fishing Brook site were about 2 orders of magnitude higher than that of wet deposition. Annual stream yields at Fishing Brook ranged from 30.9 to 45.5 (kg/ha)/yr and had a mean of 39.1 (kg/ha)/yr (fig. 14; table 11). Based on the mean annual estimates, only 1.9 percent of the chloride stream yield can be attributed to atmospheric wet deposition of chloride in the Fishing Brook basin (table 11). Annual chloride stream yields at both basins may have contributions from a number of non-precipitation sources, including evaporation-transpiration processes, weathering of bedrock, and groundwater discharge, and, at the Fishing Brook site, road salt application.

In contrast to chloride, sulfate can be involved in common aqueous geochemical (biotic and abiotic) reactions; therefore, sulfate is not considered to be a conservative element. Annual atmospheric wet deposition of sulfate at the NADP SC06 site near the McTier Creek site ranged from 9.23 to 14.9 (kg/ha)/yr and had a mean of 12.6 (kg/ha)/yr (fig. 13; table 11). Annual sulfate stream yields at the McTier Creek site were about half of the wet deposition and ranged from 3.22 to 7.30 (kg/ha)/yr, with a mean of 5.26 (kg/ha)/yr (fig. 14; table 11). Based on the mean annual estimates, about 42 percent of the atmospheric wet deposition of sulfate reaches the stream at the McTier Creek site. Annual atmospheric wet deposition of sulfate at the NADP NY20 site near Fishing Brook ranged from 8.23 to 15.1 (kg/ha)/yr and had a mean of 11.9 (kg/ha)/yr (fig. 14; table 11). Mean annual stream yield of sulfate was 29.1 (kg/ha)/yr at the Fishing Brook site and was more than twice as high as the mean annual sulfate wet deposition of 11.9 (kg/ha)/yr for the basin (table 11). Based on mean annual estimates, only 41 percent of the sulfate stream yield can be attributed to atmospheric wet deposition of chloride in the Fishing Brook basin (table 11). Annual sulfate stream yields may have contributions from a number of non-precipitation sources, including evaporation-transpiration processes, biologically mediated geochemical reactions, and weathering of bedrock and groundwater discharge.

Table 10. Estimated annual total (filtered plus particulate) total mercury stream yields and annual total mercury wet deposition from litterfall and precipitation in the McTier Creek near New Holland, S.C. (station 02172305) and Fishing Brook at County Line Flow near Newcomb, N.Y. (station 0131199050) basins, water years 2005 to 2009 and at the Mercury Deposition Network (MDN) sites at Cape Romain National Wildlife Refuge in South Carolina (SC05) and Biscuit Brook in New York (NY68), water years 2008 and 2009.

[Annual total mercury wet deposition from precipitation from PRISM models of nearby MDN sites SC19 and NY20 and from Risch and others (2012) for MDN sites SC05 and NY68. —, no data; (µg/m²/yr, micrograms per square meter per year; a water year extends from October of one calendar year through September of the next calendar year]

Estimates for McTier Creek near New Holland, S.C. / Cape Romain National Wildlife Refuge, Charleston County, S.C., at MDN site SC05 (Risch and others, 2012)

Water year	PRISM-derived annual total mercury wet deposition [(µg/m²/yr)]	Annual total mercury yield in stream [(µg/m²/yr)]	Annual total mercury deposition from litterfall [(µg/m²/yr)]	Annual total mercury deposition [(µg/m²/yr)]	Percent of wet deposition of total deposition	Percent of wet deposition that reaches stream	Percent of total deposition that reaches stream	Annual total mercury wet deposition [(µg/m²/yr)]	Annual total mercury deposition from litterfall [(µg/m²/yr)]	Annual total mercury deposition [(µg/m²/yr)]	Percent of wet deposition of total deposition
2005	11.0	2.09	—	—	—	19	—	—	—	—	—
2006	9.85	1.49	—	—	—	15	—	—	—	—	—
2007	7.60	1.37	12.8	20.4	37.3	18	7	—	—	—	—
2008	10.3	0.91	—	—	—	9	—	9.0	9.1	18.1	49.7
2009	12.2	1.44	—	—	—	12	—	7.4	9.4	16.8	44.0
Mean	9.91	1.46	—	—	—	15	—	8.2	9.3	17.5	46.9

Estimates for Fishing Brook at County Line Flow near Newcomb, N.Y. / Biscuit Brook, Ulster County, N.Y., MDN site NY68 (Risch and others, 2012)

Water year	PRISM-derived annual total mercury wet deposition [(µg/m²/yr)]	Annual total mercury yield in stream [(µg/m²/yr)]	Annual total mercury deposition from litterfall [(µg/m²/yr)]	Annual total mercury deposition [(µg/m²/yr)]	Percent of wet deposition of total deposition	Percent of wet deposition that reaches stream	Percent of total deposition that reaches stream	Annual total mercury wet deposition [(µg/m²/yr)]	Annual total mercury deposition from litterfall [(µg/m²/yr)]	Annual total mercury deposition [(µg/m²/yr)]	Percent of wet deposition of total deposition
2005	6.68	1.52	—	—	—	23	—	—	—	—	—
2006	6.96	2.30	—	—	—	33	—	—	—	—	—
2007	6.40	1.94	9.00	15.4	41.6	30	13	—	—	—	—
2008	7.17	2.11	—	—	—	29	—	10.6	16.0	26.6	39.8
2009	6.24	1.84	—	—	—	29	—	8.5	14.6	23.1	36.8
Mean	6.30	1.94	—	—	—	31	—	9.6	15.3	24.9	38.6

Table 11. Estimated annual wet deposition and stream yields of chloride and sulfate in the McTier Creek near New Holland, S.C. (02172305) and Fishing Brook at County Line Flow near Newcomb, N.Y. (0131199050) basins, water years 2005 to 2009.

[(kg/ha)/yr; kilograms per hectare per year; Wet deposition data from monitoring at the National Atmospheric Deposition Program National Trend Network (NADP NTN) site SC06 in South Carolina and based on monitoring at the NADP NTN site NY20 in New York for water years 2005 to 2009 (see table 2 for site details or access *http://nadp. sws.uiuc.edu/data/ntndata.aspx*)]

	McTier Creek					
Water year	Atmospheric wet deposition of chloride [(kg/ha)/yr]	Stream yields of chloride [(kg/ha)/yr]	Percent of chloride in stream attributed to wet deposition	Atmospheric wet deposition of sulfate [(kg/ha)/yr]	Stream yields of sulfate [(kg/ha)/yr]	Percent of wet deposition that reaches stream
2005	2.48	12.4	20	11.7	7.30	62
2006	4.37	10.2	43	14.7	5.50	38
2007	4.14	9.06	46	12.3	5.03	41
2008	3.54	6.98	51	14.9	3.22	22
2009	4.01	9.31	43	9.23	5.22	57
Mean	3.71	9.61	39	12.6	5.26	42
	Fishing Brook					
Water year	Atmospheric wet deposition of chloride [(kg/ha)/yr]	Stream yields of chloride [(kg/ha)/yr]	Percent of chloride in stream attributed to wet deposition	Atmospheric wet deposition of sulfate [(kg/ha)/yr]	Stream yields of sulfate [(kg/ha)/yr]	Percent of sulfate in stream attributed to wet deposition
2005	0.72	30.9	2.3	11.3	22.9	50
2006	1.2	45.5	2.6	12.2	32.9	37
2007	0.61	38.9	1.6	12.5	29.5	42
2008	0.66	41.5	1.6	15.1	31.5	48
2009	0.56	38.9	1.4	8.23	28.9	28
Mean	0.75	39.1	1.9	11.9	29.1	41

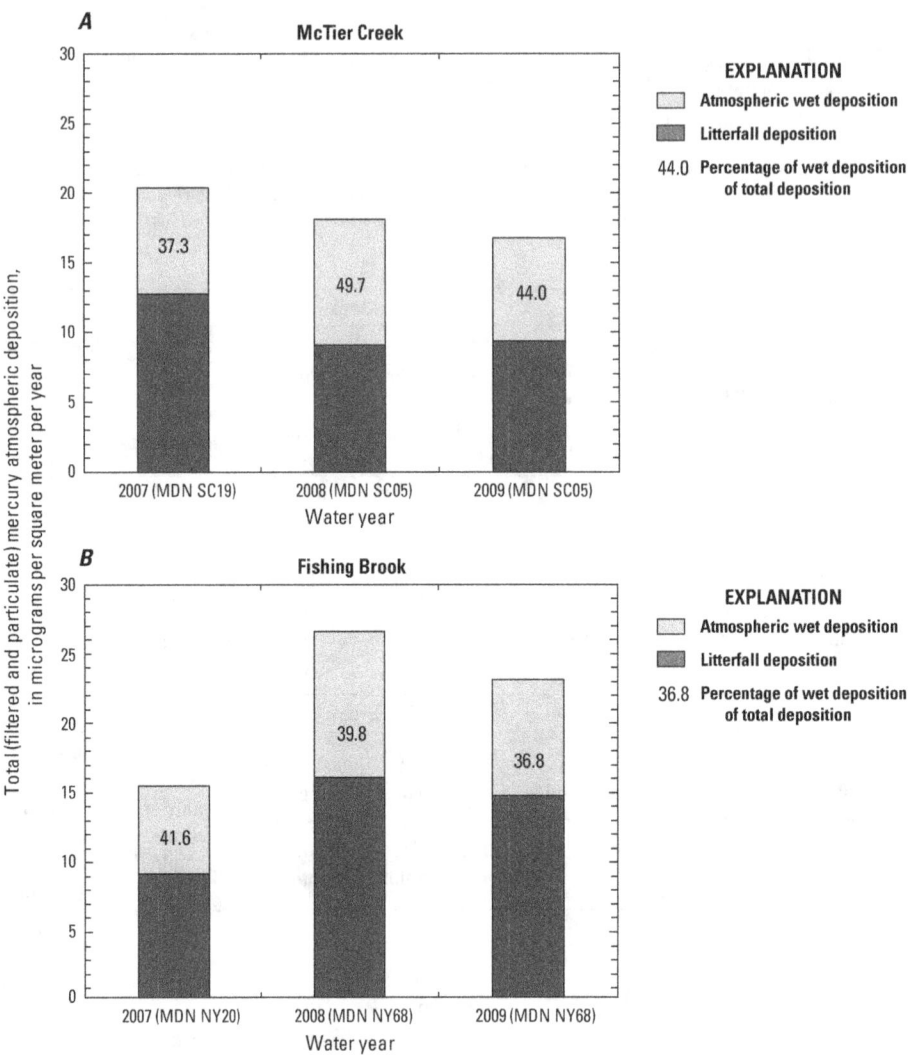

Figure 13. Estimated annual total (filtered plus particulate) mercury wet deposition from precipitation and litterfall deposition in the (*A*) McTier Creek near New Holland, S.C. (02172305) and (*B*) Fishing Brook at County Line Flow near Newcomb, N.Y. (0131199050) basins, water years 2007 to 2009. [PRISM-derived estimates of wet deposition were obtained from Congaree Swamp Mercury Deposition Network (MDN) site SC19 and from Huntington Forest MDN site NY20 for 2005 to 2009 (see tables 2 and 10). Atmospheric wet deposition and litterfall deposition were estimated for 2008 and 2009 for Cape Romain National Wildlife Refuge MDN site SC05 and for Biscuit Brook MDN Site NY68 (table 10; Risch and others, 2012)]

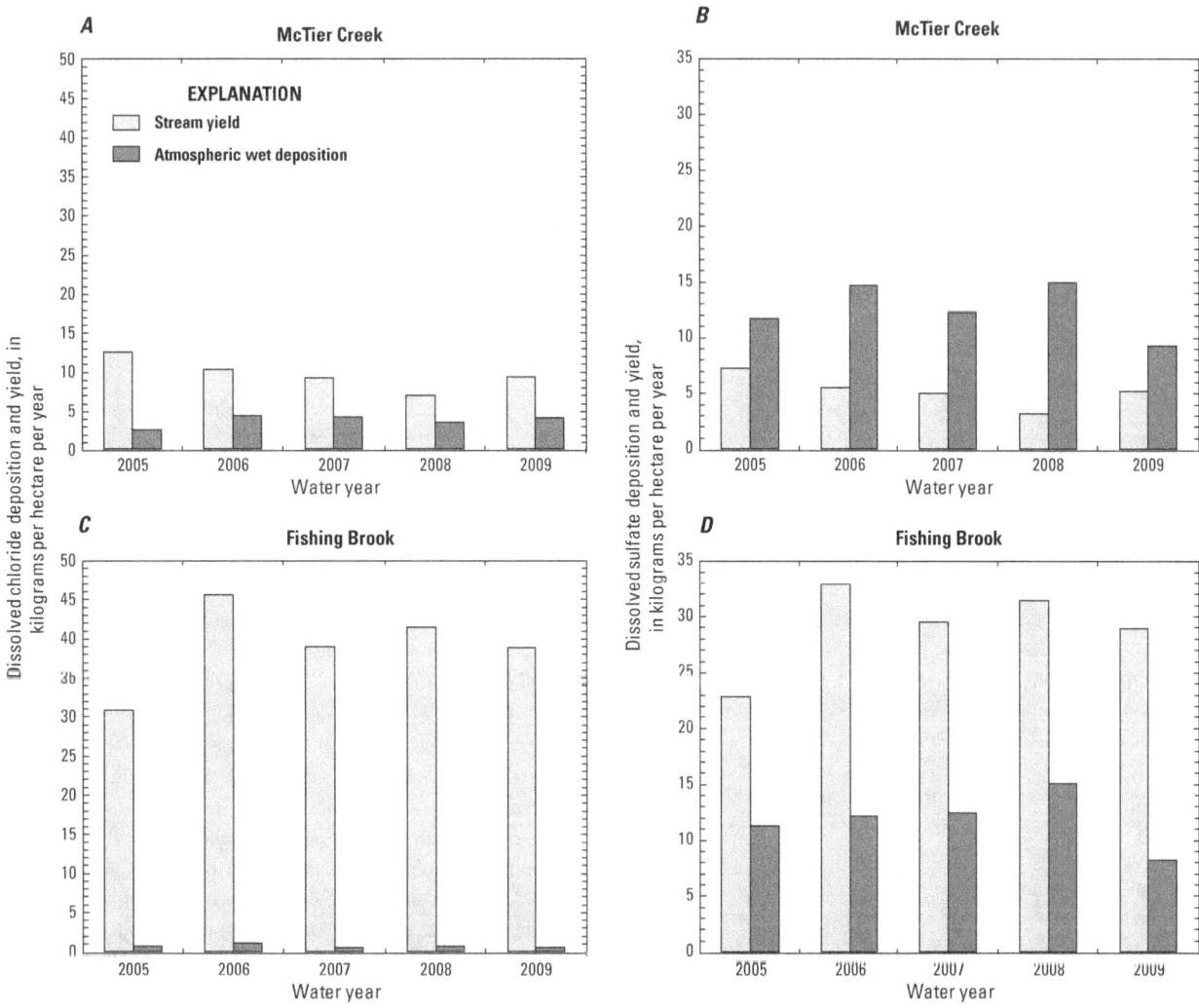

Figure 14. Estimated annual dissolved chloride and sulfate stream yields and atmospheric wet deposition from precipitation in the (*A, B*) McTier Creek near New Holland, S.C. (02172305) and (*C, D*) Fishing Brook at County Line Flow near Newcomb, N.Y. (0131199050) basins, water years 2005 to 2009.

Summary

As part of the Mercury Topical Study of the U.S. Geological Survey National Water-Quality Program, a spatially extensive assessment of the environmental controls on Hg transport and bioaccumulation in stream ecosystems was conducted in New York and South Carolina from 2005 to 2009. The study design paired a headwater stream with its larger river basin. In the Coastal Plain region of South Carolina, the study area included McTier Creek near New Holland (USGS station 02172305), which is a tributary to the Edisto River, and the Edisto River near Givhans (USGS station 02175000). In the Adirondack region of New York, the study area included the headwater site, Fishing Brook near Newcomb (USGS station 0131199050), which is a tributary to the upper Hudson River, and the Hudson River near Newcomb (USGS station 01312000). The purpose of this report is to describe and compare the concentrations, loads, and yields of particulate methylmercury (PMeHg), filtered methylmercury (FMeHg), particulate total mercury (PTHg), filtered total mercury (FTHg), suspended sediment (SSC), dissolved organic carbon (DOC), particulate organic carbon (POC), and selected major ions in the Edisto River basin at the McTier Creek and Edisto River sites and in the upper Hudson River basin at the Fishing Brook and Hudson River sites.

For the Edisto River site, annual mean streamflow for the study period was consistently below the long-term (1940 to 2009) mean annual streamflow of 70.3 cubic meters per second (m^3/s), indicating low-flow conditions. For the Hudson River, annual mean streamflow for the study period was consistently above the long-term (1925 to 2009) mean annual streamflow of 11.6 m^3/s, indicating relatively high-flow conditions.

Median concentrations of FTHg ranged from 1.55 nanograms per liter (ng/L) at the Hudson River site to 2.77 ng/L at the Edisto River site. The Edisto River site had the greatest median FMeHg concentration of 0.32 ng/L, and the Hudson River site had the least median FMeHg concentration of 0.07 ng/L. The McTier Creek and Edisto River sites had median PTHg concentrations of 0.99 and 1.19 ng/L, respectively, which were more than 2 times larger than those of the Fishing Brook and Hudson River sites (0.40 and 0.30 ng/L, respectively). Median PMeHg concentrations ranged from 0.01 ng/L at the Hudson River site to 0.05 ng/L at the Edisto River site.

At the McTier Creek site, particulate and filtered concentrations of THg were positively correlated with streamflow, indicating an increased contribution during high-flow conditions. However, FMeHg concentrations were negatively correlated with streamflow, indicating dilution during high-flow conditions. Streamflow was not correlated significantly with PMeHg concentrations, but was weakly, positively correlated to PTHg concentrations at the McTier Creek site. At the Edisto River site, FTHg and FMeHg concentrations were correlated positively with streamflow, but PTHg and PMeHg concentrations were not correlated. At the McTier Creek and Edisto River sites, streamflow and DOC concentrations had statistically significant positive correlations similar to concentrations of FTHg. As was observed with the particulate forms of Hg, streamflow was not correlated significantly to SSC at the McTier Creek and Edisto River sites.

Annual yields, which are the annual loads divided by drainage basin size, were used to compare contributions between headwater and large river sites in each basin. The 2-year (2008 and 2009) mean annual FMeHg yield of 0.025 microgram per square meter per year [(μg/m^2)/yr] at the McTier Creek site was almost 3 times smaller than the FMeHg yield of 0.064 (μg/m^2)/yr at the Edisto River site, indicative of increased contributions with increasing scale from headwater stream catchment to larger river basin. Two-year mean annual yields of FTHg were 0.83 and 0.68 (μg/m^2)/yr at the McTier and Edisto sites, respectively, and are indicative of negligible change in contribution from headwater stream to larger river scale. Yields of particulate forms of mercury were relatively consistent between headwater stream and larger river basin scales. Two-year mean annual yields of PMeHg were 0.009 and 0.011 (μg/m^2)/yr at the McTier Creek and Edisto River sites, respectively. Annual PTHg yields were 0.34 (μg/m^2)/yr at the McTier Creek site and 0.27 (μg/m^2)/yr at the Edisto River site.

In contrast to the FMeHg yields at the South Carolina sites, the FMeHg yield of 0.095 (μg/m^2)/yr at the Fishing Brook site in New York was higher than the FMeHg yield of 0.068 (μg/m^2)/yr at the Hudson River site, indicating decreased contributions with increasing scale from headwater stream catchment to larger river basin. As observed with the South Carolina sites, the 2-year mean yields of FTHg indicated no change in contribution from headwater stream to larger river scale and were 1.67 and 1.66 (μg/m^2)/yr at the Fishing Brook and Hudson River sites, respectively. Two-year mean annual yields of PMeHg were 0.016 and 0.018 (μg/m^2)/yr at the Fishing Brook and Hudson River sites, respectively, but the 2-year mean annual yield of PTHg at the Fishing Brook site of 0.30 (μg/m^2)/yr was 32 percent lower than the mean annual PTHg yield of 0.44 (μg/m^2)/yr at the Hudson River site.

The 5-year mean annual FTHg yields of 1.01 and 0.754 (μg/m^2)/yr at the McTier Creek and Edisto River paired basin sites, respectively, were lower than the mean annual FTHg yields of 1.64 and 1.60 (μg/m^2)/yr at the Fishing Brook and Hudson River paired basin sites, respectively. The McTier Creek site also had a mean annual FMeHg yield of 0.030 (μg/m^2)/yr that was 2 to 3 times lower than the mean annual FMeHg yields of 0.073 (μg/m^2)/yr at the Edisto River site, 0.092 (μg/m^2)/yr at the Fishing Brook site, and 0.063 (μg/m^2)/yr at the Hudson River site. Particulate species of mercury had lower yields than filtered species at all sites, and yields were relatively similar among the four sites. Mean annual PTHg yields ranged from 0.288 (μg/m^2)/yr (Edisto River) to 0.454 (μg/m^2)/yr (McTier Creek). Mean annual PMeHg yields ranged from 0.012 (μg/m^2)/yr (McTier Creek, Edisto River) to 0.017 (μg/m^2)/yr (Hudson River).

Relatively similar mean annual DOC yields were observed at the paired basin sites in New York and South Carolina; however, differences were observed between basins. Mean annual DOC yields of 24.4 and 16.7 kilograms per hectare per year [(kg/ha)/yr] at the McTier Creek and Edisto River paired basin sites, respectively, were lower than the mean annual DOC yields of 54.4 and 52.9 (kg/ha)/yr at the Fishing Brook and Hudson River paired basin sites, respectively. Mean annual yields of dissolved chloride and sulfate also were lower at the McTier Creek and Edisto River paired basin sites when compared to those of the Fishing Brook and Hudson River paired basin sites. Mean annual SSC yields did not exhibit a consistent pattern between the New York and South Carolina basins or between paired (headwater-large river) sites.

Total (filtered plus particulate) Hg, dissolved chloride, and dissolved sulfate yields in the streams at the study sites were compared to Mercury Deposition Network (mercury) and National Trends Network (chloride and sulfate) wet atmospheric deposition and litterfall (THg only) to determine how much of the atmospheric deposition reaches the stream. Annual THg deposition from litterfall also was computed for water year 2007 in both basins to provide an estimate of the dry deposition of THg (a water year extends from October of one calendar year to September of the next calendar year).

In the McTier Creek headwater stream basin, the PRISM-estimated annual wet deposition of THg ranged from 7.60 to 12.2 (μg/m^2)/yr and averaged 9.91 (μg/m^2)/yr. Litterfall accounted for an estimated 12.8 (μg/m^2)/yr of THg in this basin in 2007, which produced a total (wet plus litterfall) THg deposition of 20.4 (μg/m^2)/yr. Based on these estimates, wet deposition of THg represented only 37 percent of the total THg deposition in the McTier Creek basin in 2007. At the McTier Creek site, annual THg (filtered plus particulate) stream yields represented 9 to 19 percent of the estimated annual wet deposition of THg in the basin. In 2007, only 7 percent of the total (litterfall deposition plus wet deposition) THg deposition in the McTier Creek basin reached the stream site, indicating storage of atmospherically derived THg within the basin.

The Fishing Brook headwater stream basin had lower PRISM-estimated annual wet deposition of THg than the McTier Creek site. The annual wet deposition of THg ranged from 6.24 to 7.17 (μg/m^2)/yr and averaged 6.30 (μg/m^2)/yr for water years 2005 to 2009 in this basin. Litterfall accounted for an estimated 9.00 (μg/m^2)/yr of THg in this basin in 2007, which produced a total THg deposition of 15.4 (μg/m^2)/yr. Based on these estimates, wet deposition of THg represented only 42 percent of the total THg deposition in the Fishing Brook basin in 2007. Annual THg yields in the stream at the Fishing Brook site represented 23 to 33 percent of the estimated THg being delivered to the basin annually by wet deposition. In 2007, only 13 percent of the total (litterfall deposition plus wet deposition) THg deposition in the Fishing Brook basin reached the stream site, indicating storage of the atmospherically derived THg within the basin.

References

Aiken, G.R., McKnight, D.M., Thorn, K.A., and Thurman, E.M., 1992, Isolation of hydrophilic organic acids from water using nonionic macroporous resins: Organic Geochemistry, v. 18, p. 567–573.

Akaike, H., 1974, A new look at statistical model identification: IEEE Transactions on Automatic Control AU-19, p. 716–722.

American Public Health Association, 1995a, UV-absorbing organic constituents, nineteenth edition of standard methods for the examination of water and wastewater, p. 5-60 to 5-62.

American Public Health Association, 1995b, Determination of biomass (standing crop), nineteenth edition of standard methods for the examination of water and wastewater, p. 10–25.

American Public Health Association, 1998, Standard methods for the examination of water and wastewater (20th ed.): Washington, D.C., American Public Health Association, American Water Works Association, and Water Environment Federation, p. 3-37 to 3-43.

Arar, E.J., and Collins, G.B., 1997, U.S. Environmental Protection Agency Method 445.0, *In vitro* determination of chlorophyll *a* and pheophytin *a* in marine and freshwater algae by fluorescence, revision 1.2: Cincinnati, Ohio, U.S. Environmental Protection Agency, National Exposure Research Laboratory, Office of Research and Development.

Babiarz, C.L., Hurley, J.P., Hoffmann, S.R., Andren, A.W., Shafer, M.M., and Armstrong, D.E., 2001, Partitioning of total mercury and methylmercury to the colloidal phase in fresh waters: Environmental Science & Technology, v. 35, no. 24, p. 4773–4782.

Bauch, N.J., Chasar, L.C., Scudder, B.C., Moran, P.W., Hitt, K.J., Brigham, M.E., Lutz, M.A., and Wentz, D.A., 2009, Data on mercury in water, streambed sediment, and fish tissue from selected streams across the United States, 1998–2005: U.S. Geological Survey Data Series 307, 33 p.

Bischoff, J.M., Bukaveckas, P., and Mitchell, M.J., 2001, N storage and cycling in vegetation of a forested wetland—Implications for watershed N processing: Water, Air, & Soil Pollution, v. 128, p. 97–114.

Bloom, N., 1989, Determination of picogram levels of methylmercury by aqueous phase ethylation, followed by cryogenic gas chromatography with cold vapour atomic fluorescence detection: Canadian Journal of Fisheries and Aquatic Sciences, v. 46, no. 7, p. 1131–1140.

Bradley, P.M., Burns, D.A., Murray, K.R., Brigham, M.E., Button, D.T., Chasar, L.C., Di-Pasquale, M.M., Lowery, M.A., and Journey, C.A., 2011, Spatial and seasonal variability of dissolved methylmercury in two stream basins in the Eastern United States: Environmental Science & Technology, v. 45, p. 2048–2055.

Bradley, P.M., Chapelle, F.H., and Journey, C.A., 2009, Comparison of methylmercury production and accumulation in sediments of the Congaree and Edisto River basins, South Carolina, 2004–06: U.S. Geological Survey Scientific Investigations Report 2009–5021, 9 p.

Bradley, P.M., Journey, C.A., Chapelle, F.H., Lowery, M.A., and Conrads, P.A., 2010, Flood hydrology and methylmercury availability in Coastal Plain rivers: Environmental Science & Technology, v. 44, p. 9285–9290.

Brenton, R.W., and Arnett, T.L., 1993, Methods of analysis by the U.S. Geological Survey National Water Quality Laboratory—Determination of dissolved organic carbon by uv-promoted persulfate oxidation and infrared spectrometry: U.S. Geological Survey Open-File Report 92–480, 12 p.

Brigham, M.E., Wentz, D.A., Aiken, G.R., and Krabbenhoft, D.P., 2009, Mercury cycling in stream ecosystems. 1. Water column chemistry and transport: Environmental Science & Technology, v. 43, no. 8, p. 2720–2725.

Brumbaugh, W.G., Krabbenhoft, D.P., Helsel, D.R., Wiener, J.G., and Echols, K.R., 2001, A national pilot study of mercury contamination of aquatic ecosystems along multiple gradients—Bioaccumulation in fish: U.S. Geological Survey Biological Science Report 2001–0009, 25 p.

Bushey, J.T., Nallana, A.G., Montesdeoca, M.R., and Driscoll, C.T., 2008a, Mercury dynamics of a northern hardwood canopy: Atmospheric Environment, v. 42, p. 6905–6914.

Bushey, J.T., Driscoll, C.T., Mitchell, M.J., Selvendiran, P., and Montesdeoca, M.R., 2008b, Mercury transport in response to storm events form a northern forest landscape: Hydrological Processes, doi: 10.1002/hyp.7091.

Carter, R.W., and Davidian, J., 1968, General procedure for gaging streams: U.S. Geological Survey Techniques of Water Resources Investigations, book 3, chap. A6.

Castellarin, A., Vogel, R.M., and Brath, A., 2004, A stoichastic index flow model of flow duration curves: Water Resources Research, v. 40, W030104, doi:10.1029/2003WR002524, 10 p.

Chasar, L.C., Scudder, B.C., Stewart, A.R., Bell, A.H., and Aiken, G.R., 2009, Mercury cycling in stream ecosystems. 3. Trophic dynamics and methylmercury bioaccumulation: Environmental Science & Technology, v. 43, no. 8, p. 2733–2739.

Childress, C.J.O., Foreman, W.T., Conner, B.F., and Maloney, T.J., 1999, New reporting procedures based on long-term method detection levels and some considerations for interpretations of water-quality data provided by the U.S. Geological Survey National Water Quality Laboratory: U.S. Geological Survey Open-File Report 99–193, 19 p. (Also available at *http://water.usgs.gov/owq/OFR_99-193/ofr99_193.pdf.*)

Choi, H.D., Sharac, T.J., and Holsen, T.M., 2008, Mercury deposition in the Adirondacks—A comparison between precipitation and throughfall: Atmospheric Environment, v. 42, p. 1818–1827.

Cohn, T.A., 2005, Estimating contaminant loads in rivers—An application of adjusted maximum likelihood to type 1 censored data: Water Resources Research, v. 41, no. 7, W07003.

Cohn, T.A., Delong, L.L., Gilroy, E.J., Hirsch, R.M., and Wells, D.K., 1989, Estimating constituent loads: Water Resources Research, v. 25, no. 5, p. 937–942.

Cohn, T.A., Caulder, D.L., Gilroy, E.J., Zynjuk, L.D., and Summers, R.M., 1992, The validity of a simple statistical model for estimating fluvial constituent loads—An empirical study involving nutrient loads entering Chesapeake Bay: Water Resources Research, v. 28, no. 9, p. 2353–2363.

Cooney, T.W., 2001, Surface water quality-assurance plan for the South Carolina District of the U.S. Geological Survey: U.S. Geological Survey Open-File Report 01-121, 48 p., accessed March 16, 2004, at *http://sc.water.usgs.gov/publications/abstracts/ofr01-121.html.*

Crawford, C.G., 1991, Estimation of suspended-sediment rating curves and mean suspended-sediment loads: Journal of Hydrology, v. 129, p. 331–348.

Daly, C., 2006, Guidelines for assessing the suitability of spatial climate data sets: International Journey of Climatology, v. 26, p. 707–721.

Daly, C., Neilson, R.P., and Phillips, D.L., 1994, A statistical-topographic model for mapping climatological precipitation over mountanious terrain: Journal of Applied Meteorology, v. 33, p. 140–158.

Daly, C., Gibson, W.P., Taylor, G.H., Johnson, G.L., and Pasteris, P., 2002, A knowledge-based approach to the statistical mapping of climate: Climate Research, v. 22, p. 99–113.

DeWild, J.F., Olson, M.L, and Olund, S.D., 2002, Determination of methyl mercury by aqueous phase ethylation, followed by gas chromatographic separation with cold vapor atomic fluorescence detection: U.S. Geological Survey Open-File Report 01–445, 14 p., accessed October 19, 2009, at *http://pubs.er.usgs.gov/usgspubs/ofr/ofr2001445.*

DeWild, J.F., Olund, S.D., Olund, M.L., and Tate, M.T., 2004, Methods for the preparation and analysis of solids and suspended solids for methylmercury: U.S. Geological Survey Techniques and Methods, book 5, chap. 7, accessed May 29, 2012, at *http://pubs.usgs.gov/tm/2005/tm5A7/*.

Dittman, J.A., Shanley, J.B., Driscoll, C.T., Aiken, G.R., Chambers, A.T., and Towse, J.E., 2009, Ultraviolet absorbance as a proxy for total dissolved mercury in streams: Environmental Pollution, v. 157, p. 1953–1956.

Dittman, J.A., Shanley, J.B., Driscoll, C.T., Aiken, G.R., Chambers, A.T., Towse, J.E., and Selvendiran, P., 2010, Mercury dynamics in relation to dissolved organic carbon concentration and quality during high flow events in three northeastern U.S. streams: Water Resources Research, v. 46, W07522, doi:10.1029/2009WR008351.

Downs, S.G., Macleod, C.L., and Lester, J.N., 1998, Mercury in precipitation and its relation to bioaccumulation in fish—A literature review: Water, Air, and Soil Pollution, v. 108, p. 149–187.

Driscoll, C.T., Han, Y.J., Chen, C.Y., Evers, D.C., Lambert, K.F., Holsen, T.M., Kamman, N.C., and Munson, R.K., 2007, Mercury contamination in forest and freshwater ecosystems in the Northeastern United States: BioScience, v. 57, p. 17–28.

Evers, D.C., Young-Ji, H., Driscoll, C.T., Kamman, N.C., Goodale, M.W., Lambert, K.F., Holsen, T.M., Chen, C.Y., Clair, T.A., and Butler, T., 2007, Biological mercury hotspots in the northeastern United States and southeastern Canada: BioScience, v. 57, no. 1, p. 29–43.

Feaster, T.D., Golden, H.E., Odom, K.R., Lowery, M.A., Conrads, P.A., and Bradley, P.M., 2010, Simulation of streamflow in the McTier Creek watershed, South Carolina: U.S. Geological Survey Scientific Investigations Report 2010–5202, 61 p.

Fishman, M.J., ed., 1993, Methods of analysis by the U.S. Geological Survey National Water Quality Laboratory—Determination of inorganic and organic constituents in water and fluvial sediments: U.S. Geological Survey Open-File Report 93–125, 217 p.

Fishman, M.J., and Friedman, L.C., 1989, Methods for determination of inorganic substances in water and fluvial sediments: U.S. Geological Survey Techniques of Water-Resources Investigations, book 5, chap. A1, 545 p.

Fitzgerald, W.F., and Watras, C.J., 1989, Mercury in surficial waters of rural Wisconsin lakes: Science of the Total Environment, v. 87–88, p. 223–232.

Fry, J.A., Coan, M.J., Homer, C.G., Meyer, D.K., and Wickham, J.D., 2009, Completion of the National Land Cover Database (NLCD) 1992–2001 Land Cover Change Retrofit product: U.S. Geological Survey Open-File Report 2008–1379, 18 p.

Gabriel, M.C., and Williamson, D.G., 2004, Principal biogeochemical factors affecting the speciation and transport of mercury through the terrestrial environment: Environmental Geochemistry and Health, v. 26, p. 421–434.

Glover, J.B., Domino, M.E., Altman, K.C., Dillman, J.W., Castleberry, W.S., Eidson, J.P., and Mattocks, M., 2010, Mercury in South Carolina fishes, USA: Ecotoxicology, v. 19, no. 4, p. 781–795.

Griffith, G.E., Omernik, J.M., Comstock, J.A., Schafale, M.P., McNab, W.H., Lenat, D.R., MacPherson, T.F., Glover, J.B., and Shelburne, V.B., 2002, Ecoregions of North Carolina and South Carolina: U.S. Environmental Protection Agency Map (1:1,500,000 scale), accessed April 13, 2010, at *http://www.epa.gov/wed/pages/ecoregions/ncsc.htm*.

Grigal, D.F., 2002, Inputs and outputs of mercury from terrestrial watersheds—A review: Environmental Reviews, v. 10, p. 1–39.

Grigal, D.F., 2003, Mercury sequestration in forests and peatlands—A review: Journal of Environmental Quality, v. 32, p. 292–405.

Haitzer, M., Aiken, G.R., and Ryan, J.N., 2002, Binding of mercury(II) to dissolved organic matter—The role of the mercury-to-DOM concentration ratio: Environmental Science & Technology, v. 36, p. 3564–3570.

Hammerschmidt, C.R., and Fitzgerald, W.R., 2006, Bioaccumulation and trophic transfer of methylmercury in Long Island Sound: Archives of Environmental Contamination and Toxicology, v. 51, p. 416–424.

Helsel, D.R., 2005, Nondetects and data analysis—Statistics for censored environmental data: New Jersey, Wiley Interscience, 250 p.

Helsel, D.R., and Hirsch, R.M., 1992, Statistical methods in water sources. U.S. Geological Survey Techniques of Water Resources Investigations, book 4, chap. A3, 512 p.

Herczeg, A.L., and Edmunds, W.M., 2000, Inorganic ions as tracers, *in* Cook, P.G., and Herczeg, A.L., eds., Environmental tracers in subsurface hydrology: Boston, Kluwer Academic Publishers, p. 31–77.

Hirsch, R.M., 1982, A comparison of four streamflow record extension techniques: Water Resources Research, v. 18, no. 4, p. 1081–1088.

Homer, C., Huang, C., Yang, L., Wylie, B., and Coan, M., 2004, Development of a 2001 National Landcover Database for the United States: Photogrammetric Engineering and Remote Sensing, v. 70, no. 7, p. 829–840.

Horvat, M., Liang, L., and Bloom, N.S., 1993, Comparison of distillation with other current isolation methods for the determination of methyl mercury compounds in low-level environmental samples—Part II. Water: Analytica Chemica Acta, v. 282, no. 1, p. 153–168.

Hurley, J.P., Benoit, J.M., Babiaz, C.L., Shafer, M.M., Andren, A.W., Sullivan, J.R., Hammond, R., and Webb, D.A., 1995, Influences of watershed characteristics on mercury levels in Wisconsin Rivers: Environmental Science & Technology, v. 29, p. 1867–1875.

Hurley, J.P., Krabbenhoft, D.P., Cleckner, L.B., Olson, M.L., Aiken, G.R., and Rawlik, P.S., Jr., 1998, System controls on the aqueous distribution of mercury in the northern Florida Everglades: Biogeochemistry, v. 40, p. 293–310.

Jackson, T.A., 1997, Long-range atmospheric transport of mercury to ecosystems, and the importance of anthropogenic emissions—A critical review and evaluation of the published evidence: Environmental Review, v. 5, p. 99–120.

Jain, S.K., and Sudheer, K.P., 2008, Fitting of hydrologic models—A close look at the Nash-Sutcliffe Index: Journal of Hydrologic Engineering, v. 13, no. 10, p. 981–986.

Johnson, D.W., and Lindberg, S.E., 1995, The biogeochemical cycling of Hg in forests—Alternative methods for quantifying total deposition and soil emission: Water, Air, and Soil Pollution, v. 80, p. 1069–1077.

Kennedy, E.J., 1984, Discharge ratings at gaging stations: U.S. Geological Survey Techniques of Water Resources Investigations, book 3, chap. A10.

Kolka, R.K., Grigal, D.F., Verry, E.S., and Nater, E.A., 1999, Mercury and organic carbon relationships in streams draining forested upland/peatland watersheds: Journal of Environmental Quality, v. 28, p. 766–775.

Latysh, N.E., and Wetherbee, G.A., 2007, External quality assurance programs managed by the U.S. Geological Survey in support of the National Atmospheric Deposition Program/Mercury Deposition Network: U.S. Geological Survey Open-File Report 2007–1170.

Latysh, N.E., and Wetherbee, G.A., 2011, Improved mapping of National Atmospheric Deposition Program wet-deposition in complex terrain using PRISM-gridded data sets: Environmental Monitoring and Assessment: doi10.1007/s10661-011-209-7, accessed July 5, 2011, at *http://www.springerlink.com/content/bg0vg56668j360j6/fulltext.pdf.*

Lewis, M.E., and Brigham, M.E., 2004, Low-level mercury: U.S. Geological Survey Techniques of Water-Resources Investigations, book 9, chap. A5, section 5.6.4.B, 26 p., accessed March 16, 2007, at *http://water.usgs.gov/owq/FieldManual/chapter5/pdf/5.6.4.B_v1.0.pdf.*

Lindqvist, O., 1991, Mercury in the Swedish environment: Water, Air, and Soil Pollution, v. 55, p. 23–32.

Marvin-DiPasquale, Mark, Lutz, M.A., Brigham, M.E., Krabbenhoft, D.P., Aiken, G.R., Orem, W.H., and Hall, B.D., 2009, Mercury cycling in stream ecosystems. 2. Benthic methylmercury production and bed sediment—pore water partitioning: Environmental Science & Technology, v. 43, no. 8, p. 2726–2732.

Mercury Deposition Network, 2006a, MDN sites: National Atmospheric Deposition Program, accessed November 10, 2010, at *http://nadp.sws.uiuc.edu/sites/mdnmap.asp.*

Mercury Deposition Network, 2006b, Quality assurance plan—Mercury Analytical Laboratory, 2006, accessed August 23, 2011, at *http://nadp.sws.uiuc.edu/lib/qaplans/qapHAL2006.pdf.*

Miller, E.K., Vanarsdale, A., Keeler, G.J., Chalmers, A., Poissant, L., Kamman, N.C., and Brulotte, R., 2005, Estimation and mapping of wet and dry mercury deposition across northeastern North America: Ecotoxicology, v. 14, p. 53–70.

Munthe, J., Hellsten, S., and Zetterberg, T., 2007, Mobilization of mercury and methylmercury from forest soils after a severe storm-fell event: Ambio, v. 36, no. 1, p. 111–113.

Nash, J.E., and Sutcliffe, J.V., 1970, River flow forecasting through conceptual models part I—A discussion of principles: Journal of Hydrology, v. 10, no. 3, p. 282–290.

National Atmospheric Deposition Program, 2011, NADP Network Quality Assurance Plan Version 5.1: National Atmospheric Deposition Program, accessed March 8, 2012, at *http://nadp.sws.uiuc.edu/lib/qaplans/NADP_Network_Quality_Assurance_Plan.pdf.*

National Climatic Data Center, 2011, Local climatological data publication: National Oceanic and Atmospheric Administration, National Environmental Satellite, Data, and Information Service, accessed March 30, 2011, at *http://www.ncdc.noaa.gov/oa/ncdc.html.*

National Research Council, 2000, Toxicological effects of methylmercury: Washington, D.C., National Academy Press, Committee on the Toxicological Effects of Methylmercury, Board of Environmental Studies and Toxicology, 344 p.

Olund, S.D., DeWild, J.F., Olson, M.L., and Tate, M.T., 2004, Methods for the preparation and analysis of solids and suspended solids for total mercury: U.S. Geological Survey Techniques of Water-Resources Investigations, book 5, chap. A8, accessed October 19, 2009, at *http://pubs.usgs. gov/tm/2005/tm5A8/*.

Omernik, J.M., 1987, Ecoregions of the conterminous United States, map (scale 1:7,500,000): Annals of the Association of American Geographers, v. 77, no. 1, p. 118–125.

Omernik, J.M., 2005, Ecoregions of the continental United States (Omernik Level III): U.S. Geological Survey, accessed March 15, 2008, at *http://nationalatlas.gov/mld/ ecomrpi.html*.

Patton, C.J., and Kryskalla, J.R., 2003, Methods of analysis by the U.S. Geological Survey National Water Quality Laboratory—Evaluation of alkaline persulfate digestion as an alternative to Kjeldahl digestion for determination of total and dissolved nitrogen and phosphorus in water: U.S. Geological Survey Water-Resources Investigations Report 03-4174, 33 p.

Patton, C.J., and Truitt, E.P., 1992, Methods of analysis by the U.S. Geological Survey National Water Quality Laboratory—Determination of total phosphorus by a Kjeldahl digestion method and an automated colorimetric finish that includes dialysis: U.S. Geological Survey Open-File Report 92–146, 39 p.

Patton, C.J., and Truitt, E.P., 2000, Methods of analysis by the U.S. Geological Survey National Water Quality Laboratory—Determination of ammonium plus organic nitrogen by a Kjeldahl digestion method and an automated photometric finish that includes digest cleanup by gas diffusion: U.S. Geological Survey Open-File Report 00–170, 31 p.

Price, C.V., Nakagaki, N., Hitt, K.J., and Clawges, R.M., 2007, Enhanced historical land-use and land-cover datasets of the U.S. Geological Survey: U.S. Geological Survey Data Series 240, accessed March 15, 2008, at *http://pubs.usgs. gov/ds/2006/240*.

PRISM Climate Group, 2010, PRISM Climate Group: Oregon State University, accessed June 11, 2012, at *http://prism. oregonstate.edu*.

Rantz, S.E., and others, 1982, Measurement and computation of streamflow—Volume 1. Measurement of stage and discharge: U.S. Geological Survey Water-Supply Paper 2175, 284 p.

Risch, M.R., DeWild, J.F., Krabbenhoft, D.P., Kolka, R.K., and Zhang, L., 2012, Litterfall mercury dry deposition in the eastern USA: Environmental Pollution, v. 161, p. 284–290.

Riva-Murray, K., Chasar, L.C., Bradley, P.M., Burns, D.A., Brigham, M.E., Smith, M.J., and Abrahamson, T.A., 2011, Spatial patterns of mercury in macroinvertebrates and fishes from streams of two contrasting forested landscapes in the eastern United States: Ecotoxicology, v. 20, no. 7, p. 1530–1542.

Runkel, R.L., Crawford, C.G., and Cohn, T.A., 2004, Load Estimator (LOADEST)—A FORTRAN program for estimating constituent loads in streams and rivers: U.S. Geological Survey Techniques and Methods, book 4, chap. A5, 69 p.

St. Louis, V.L., Rudd, J.W.M., Kelly, C.A., Hall, B.D., Rolfhus, K.R., Scott, K.J., Lindberg, S., and Dong, W., 2001, Importance of the forest canopy to fluxes of methyl mercury and total mercury to boreal ecosystems: Environmental Science & Technology, v. 35, p. 3089–3098.

Schelker, J., Burns, D.A., Weiler, M., and Laudon, H., 2011, Hydrological mobilization of mercury and dissolved organic carbon in a snow-dominated, forested watershed—Conceptualization and modeling: Journal of Geophysical Research, v. 116, G01002, accessed June 11, 2012, at *http://www.agu. org/pubs/crossref/2011/2010JG001330.shtml*.

Scherbatskoy, T., Shanley, J.B., and Keeler, G., 1998, Factors controlling mercury transport in an upland forested catchment: Water, Air and Soil Pollution, v. 105, p. 427–438.

Scudder, B.C., Chasar, L.C., Wentz, D.A., Bauch, N.J., Brigham, M.E., Moran, P.W., and Krabbenhoft, D.P., 2009, Mercury in fish, bed sediment, and water from streams across the United States, 1998–2005: U.S. Geological Survey Scientific Investigations Report 2009–5109, 74 p.

Scudder Eikenberry, B.C., Riva-Murray, K., Smith, M.J., Bradley, P.M., Button, D.T., Clark, J.M., Burns, D.A., and Journey, C.A., 2011, Environmental settings of streams sampled for mercury in New York and South Carolina, 2005–09: U.S. Geological Survey Open-File Report 2011–1318, 36 p.

Searcy, J.K., 1959, Flow-duration curves: U.S. Geological Survey Water-Supply Paper 1542-A, 33 p.

Selvendiran, P., Driscoll, C.T., Bushey, J.T., and Montesdeoca, M.R., 2008a, Wetland influence on mercury fate and transport in a temperate forested watershed: Environmental Pollution, v. 154, p. 46–55.

Selvendiran, P., Driscoll, C.T., Montesdeoca, M.R., and Bushey, J.T., 2008b, Inputs, storage, and transport of total and methylmercury in two temperate forest wetlands: Journal of Geophysical Research, v. 113, G00C01, doi:10.1029/2008JG000739.

Selvendiran, P., Driscoll, C.T., Montesdeoca, M.R., Choi, H.D., and Holsen, T.M., 2009, Mercury dynamics and transport in two Adirondack lakes: Limnology and Oceanography, v. 54, no. 2, p. 413–427.

Shreve, E.A., and Downs, A.C., 2005, Quality-assurance plan for the analysis of fluvial sediment by the U.S. Geological Survey Kentucky Water Science Center Sediment Laboratory: U.S. Geological Survey Open-File Report 2005–1230, 28 p.

Simonin, H.A., Loukams, J.J., Skinner, L.C., and Roy, K.M., 2008, Environmental Pollution: v. 154, p. 107–115.

Skyllberg, U., Qian, J., Frech, W., Xia, K., and Bleam, W.F., 2003, Distribution of mercury, methylmercury, and organic sulphur species in soil, soil solution and stream of a boreal forest catchment: Biogeochemistry, v. 64, p. 53–76.

Slack, J.R., Lorenz, D.L., and others, 2003, USGS library for S-PLUS for Windows—Release 2.1: U.S. Geological Survey Open-File Report 03–357.

TIBCO, 2008, TIBCO Spotfire S+® for Windows® User's Guide: TIBCO Software, Inc.

Turnipseed, D.P., and Sauer, V.B., 2010, Discharge measurements at gaging stations: U.S. Geological Survey Techniques and Methods, book 3, chap. A8, 87 p. (Also available at *http://pubs.usgs.gov/tm/tm3-a8/*.)

U.S. Environmental Protection Agency, 1996, Method 1669—Sampling ambient water for trace metals at EPA water quality criteria levels: Washington, D.C., U.S. Environmental Protection Agency Office of Water, Engineering and Analysis Division, 37 p., accessed March 12, 2011, at *http://www.epa.gov/waterscience/methods/method/inorganics/1669.pdf*.

U.S. Environmental Protection Agency, 1997, Mercury study report to Congress—Volume I. Executive summary: Office of Air Quality Planning and Standards and Office of Research and Development Report EPA452/R-97-003, 95 p.

U.S. Environmental Protection Agency, 2002, Method 1631 Revision E—Mercury in water by oxidation, purge, and trap, and cold vapor atomic fluorescence spectrometry: U.S. Environmental Protection Agency Report EPA–821–R–02–019, 38 p., available at *http://www.epa.gov/waterscience/methods/1631e.pdf*.

U.S. Environmental Protection Agency, 2005, Omernik's level III ecoregions of the continental United States: National Atlas of the United States, Reston, Va., scale 1:7,500,000, available at *http://nationalatlas.gov/mld/ecoomrp.html*.

U.S. Environmental Protection Agency, 2009, National Listing of Fish Advisories—2008 Biennial National Listing: U.S. Environmental Protection Agency Technical Fact Sheet EPA–823–F–09–007, 7 p.

U.S. Geological Survey, 2000, New method for particulate carbon and particulate nitrogen: U.S. Geological Survey Office of Water Quality Technical Memorandum 2000.08, accessed September 23, 2011, at *http://water.usgs.gov/admin/memo/QW/qw00.08.html*.

U.S. Geological Survey, variously dated, National field manual for the collection of water-quality data: U.S. Geological Survey Techniques of Water-Resources Investigations, book 9, chap. A1–A9, accessed October 19, 2009, at *http://pubs.usgs.gov/twri/*.

Vogel, R.M., and Fennessey, N.M., 1994, Flow duration curves I—A new interpretation and confidence intervals: Journal of Water Resource Planning and Management, American Society of Civil Engineers, v. 120, no. 4, p. 485–504.

Vogel, R.M., and Fennessey, N.M., 1995, Flow duration curves II—A review of applications in water resources planning: Water Resources Bulletin, v. 31, no. 6, p. 1029–1039.

Watras, C.J., Bloom, N.S., Hudson, R.J.M., Gherini, S., Munson, R., Claas, S.A., Morrison, K.A., Hurley, J., Wiener, J.G., Fitzgerald, W.F., Mason, R., Vandal, G., Powell, D., Rada, R., Rislove, L., Winfrey, M., Elder, J., Krabbenhoft, D., Andren, A.W., Babiarz, C., Porcella, D.B., and Huckabee, J.W., 1994, Sources and fates of mercury and methylmercury in Wisconsin lakes, *in* Watras, C.J., and Huckabee, J.W., eds., Mercury pollution—Integration and synthesis: Boca Raton, Fla., Lewis Publishers, p. 153–177.

Woodruff, L.G., Cannon, W.F., Knightes, C.D., Chapelle, F.H., Bradley, P.M., Burns, D.A., Brigham, M.E., and Lowery, M.A., 2011, Total mercury, methylmercury, and selected elements in soils of the Fishing Brook Watershed, Hamilton County, New York, and the McTier Creek Watershed, Aiken County, South Carolina, 2008: U.S. Geological Survey Data Series, 10 p.

Yin, X., and Balogh, S.J., 2002, Mercury concentrations in stream and river water—An analytical framework applied to Minnesota and Wisconsin (USA): Water, Air, and Soil Pollution, v. 128, p. 79–100.

Zimmermann, C.F., Keefe, C.W., and Bashe, Jerry, 1997, Method 440.0—Determination of carbon and nitrogen in sediments and particulates of estuarine/coastal waters using elemental analysis (Revision 1.4): U.S. Environmental Protection Agency, National Exposure Research Laboratory, 10 p., accessed June 11, 2012, at *http://www.epa.gov/microbes/m440_0.pdf*.

Appendixes

Appendix 1-A.   Annual mean streamflow computed for water years with complete records at McTier Creek near Monetta, S.C. (station 02172300), McTier Creek near New Holland, S.C. (station 02172305), Edisto River near Givhans, S.C. (station 02175000), Fishing Brook at County Line Flow near Newcomb, N.Y. (station 0131199050), and Hudson River near Newcomb, N.Y. (station 01312000) for the period of record at each station.

[—, no data]

Water year	Annual mean streamflow, in cubic feet per second					Water year	Annual mean streamflow, in cubic feet per second				
	McTier Creek– Monetta 02172300	McTier Creek 02172305	Edisto River 02175000	Fishing Brook 0131199050	Hudson River 01312000		McTier Creek– Monetta 02172300	McTier Creek 02172305	Edisto River 02175000	Fishing Brook 0131199050	Hudson River 01312000
1926	—	—	—	—	448.2	1968	—	—	1,518	—	348.6
1927	—	—	—	—	375.8	1969	—	—	2,733	—	463.5
1928	—	—	—	—	540.5	1970	—	—	2,319	—	344.3
1929	—	—	—	—	457	1971	—	—	3,244	—	428.6
1930	—	—	—	—	437.3	1972	—	—	3,038	—	488.3
1931	—	—	—	—	241.9	1973	—	—	4,337	—	478
1932	—	—	—	—	432.2	1974	—	—	2,466	—	434.3
1933	—	—	—	—	418.7	1975	—	—	3,186	—	400.5
1934	—	—	—	—	278.8	1976	—	—	2,477	—	585.2
1935	—	—	—	—	383	1977	—	—	2,778	—	470.4
1936	—	—	—	—	384.4	1978	—	—	2,387	—	506.7
1937	—	—	—	—	471.2	1979	—	—	3,358	—	417
1938	—	—	—	—	437.6	1980	—	—	3,311	—	373.1
1939	—	—	—	—	346.4	1981	—	—	1,326	—	399.5
1940	—	—	1,819	—	295.7	1982	—	—	1,990	—	392
1941	—	—	2,411	—	253.1	1983	—	—	2,940	—	—
1942	—	—	3,037	—	355.9	1984	—	—	3,122	—	462.3
1943	—	—	1,981	—	477.5	1985	—	—	1,377	—	414.4
1944	—	—	2,461	—	344	1986	—	—	1,978	—	489
1945	—	—	1,973	—	433.6	1987	—	—	2,424	—	384.3
1946	—	—	2,631	—	437.6	1988	—	—	1,258	—	—
1947	—	—	2,259	—	638.5	1989	—	—	1,619	—	—
1948	—	—	4,837	—	307.9	1990	—	—	2,014	—	—
1949	—	—	4,317	—	371.9	1991	—	—	4,524	—	—
1950	—	—	1,618	—	359.7	1992	—	—	2,375	—	—
1951	—	—	1,715	—	433.7	1993	—	—	3,812	—	—
1952	—	—	2,089	—	430.6	1994	—	—	1,930	—	—
1953	—	—	1,997	—	389.3	1995	—	—	3,726	—	—
1954	—	—	1,445	—	463.3	1996	26.5	—	2,440	—	—
1955	—	—	1,191	—	383	1997	21.2	—	1,648	—	—
1956	—	—	1,525	—	334.6	1998	—	—	4,839	—	—
1957	—	—	1,266	—	268.5	1999	—	—	1,407	—	—
1958	—	—	2,805	—	381.9	2000	—	—	1,560	—	—
1959	—	—	2,724	—	383.5	2001	—	—	1,380	—	—
1960	—	—	5,225	—	477.7	2002	7.18	—	645.3	—	—
1961	—	—	3,121	—	295.2	2003	23.1	—	4,201	—	408.9
1962	—	—	2,481	—	306.5	2004	17.2	—	1,621	—	543.3
1963	—	—	2,220	—	351.8	2005	18.2	—	1,765	—	413.2
1964	—	—	4,792	—	272.7	2006	14.2	—	1,193	—	603.5
1965	—	—	5,019	—	237.4	2007	12.5	—	1,306	—	532.2
1966	—	—	2,910	—	379.1	2008	9.94	19.2	929.4	61.8	603.2
1967	—	—	2,054	—	316.1	2009	12.7	26.4	1,579	55.9	498.1

Appendix 1-B. Measured and MOVE.1-estimated daily mean streamflow at McTier Creek near New Holland, S.C. (station 02172305) and Fishing Brook (County Line Flow) near Newcomb, N.Y. (station 0131199050).

[--, no data MOVE.1 regression technique used measured daily mean streamflow at long-term index station McTier Creek near Monetta, S.C. (station 02172300). MOVE.1 regression technique used measured daily mean streamflow at long-term index station Hudson River near Newcomb, N.Y. (station 01312000).]

Date	Daily mean streamflow, in cubic feet per second					
	McTier Creek			Fishing Brook		
	Measured for MOVE.1 long-term index station 02172300	Measured for 02172305	MOVE.1 estimate for 02172305	Measured at MOVE.1 long-term index station 01312000	Measured at 0131199050	MOVE.1 estimate for 013199050
10/1/2004	12	--	25.39	108.53	--	12.41
10/2/2004	12	--	25.39	106.42	--	12.17
10/3/2004	12	--	25.39	112.14	--	12.81
10/4/2004	12	--	25.39	117.24	--	13.38
10/5/2004	11	--	23.48	113.91	--	13.01
10/6/2004	11	--	23.48	108.18	--	12.37
10/7/2004	11	--	23.48	103.76	--	11.87
10/8/2004	11	--	23.48	100.04	--	11.46
10/9/2004	11	--	23.48	96.17	--	11.03
10/10/2004	11	--	23.48	92.88	--	10.66
10/11/2004	11	--	23.48	90.25	--	10.36
10/12/2004	10	--	21.56	86.50	--	9.94
10/13/2004	11	--	23.48	84.26	--	9.69
10/14/2004	11	--	23.48	81.32	--	9.36
10/15/2004	12	--	25.39	79.64	--	9.18
10/16/2004	12	--	25.39	97.88	--	11.22
10/17/2004	11	--	23.48	122.69	--	13.98
10/18/2004	9.7	--	20.98	134.58	--	15.30
10/19/2004	8	--	17.66	131.11	--	14.92
10/20/2004	8.6	--	18.84	123.59	--	14.08
10/21/2004	11	--	23.48	119.25	--	13.60
10/22/2004	12	--	25.39	117.55	--	13.41
10/23/2004	11	--	23.48	116.60	--	13.31
10/24/2004	11	--	23.48	113.28	--	12.94
10/25/2004	12	--	25.39	109.60	--	12.53
10/26/2004	12	--	25.39	104.87	--	12.00
10/27/2004	14	--	29.14	100.92	--	11.56
10/28/2004	13	--	27.27	96.72	--	11.09
10/29/2004	13	--	27.27	93.50	--	10.73
10/30/2004	13	--	27.27	91.89	--	10.55
10/31/2004	14	--	29.14	96.46	--	11.06
11/1/2004	14	--	29.14	114.27	--	13.05
11/2/2004	13	--	27.27	138.30	--	15.71
11/3/2004	13	--	27.27	182.47	--	20.59
11/4/2004	15	--	31.00	268.62	--	30.02
11/5/2004	17	--	34.68	294.42	--	32.83

Appendix 1-B.   Measured and MOVE.1-estimated daily mean streamflow at McTier Creek near New Holland, S.C. (station 02172305) and Fishing Brook (County Line Flow) near Newcomb, N.Y. (station 0131199050).—Continued

[--, no data MOVE.1 regression technique used measured daily mean streamflow at long-term index station McTier Creek near Monetta, S.C. (station 02172300). MOVE.1 regression technique used measured daily mean streamflow at long-term index station Hudson River near Newcomb, N.Y. (station 01312000).]

	Daily mean streamflow, in cubic feet per second					
	McTier Creek			Fishing Brook		
Date	Measured for MOVE.1 long-term index station 02172300	Measured for 02172305	MOVE.1 estimate for 02172305	Measured at MOVE.1 long-term index station 01312000	Measured at 0131199050	MOVE.1 estimate for 013199050
11/6/2004	14	--	29.14	321.44	--	35.76
11/7/2004	13	--	27.27	328.27	--	36.50
11/8/2004	13	--	27.27	367.33	--	40.73
11/9/2004	12	--	25.39	371.58	--	41.19
11/10/2004	12	--	25.39	314.01	--	34.96
11/11/2004	12	--	25.39	277.66	--	31.00
11/12/2004	23	--	45.45	246.99	--	27.66
11/13/2004	28	--	54.21	217.11	--	24.39
11/14/2004	17	--	34.68	194.79	--	21.94
11/15/2004	14	--	29.14	182.51	--	20.59
11/16/2004	12	--	25.39	170.45	--	19.27
11/17/2004	13	--	27.27	162.27	--	18.36
11/18/2004	14	--	29.14	156.76	--	17.76
11/19/2004	13	--	27.27	155.51	--	17.62
11/20/2004	13	--	27.27	153.89	--	17.44
11/21/2004	12	--	25.39	162.79	--	18.42
11/22/2004	13	--	27.27	177.23	--	20.01
11/23/2004	21	--	41.90	189.33	--	21.34
11/24/2004	33	--	62.80	190.11	--	21.43
11/25/2004	26	--	50.73	353.00	--	39.18
11/26/2004	18	--	36.50	1,024.01	--	110.67
11/27/2004	17	--	34.68	953.09	--	103.19
11/28/2004	24	--	47.22	719.03	--	78.40
11/29/2004	18	--	36.50	886.99	--	96.21
11/30/2004	16	--	32.84	882.22	--	95.71
12/1/2004	17	--	34.68	764.66	--	83.25
12/2/2004	16	--	32.84	962.53	--	104.19
12/3/2004	16	--	32.84	969.44	--	104.92
12/4/2004	15	--	31.00	812.12	--	88.28
12/5/2004	15	--	31.00	663.82	--	72.53
12/6/2004	15	--	31.00	513.80	--	56.50
12/7/2004	15	--	31.00	456.99	--	50.40
12/8/2004	15	--	31.00	471.13	--	51.92
12/9/2004	15	--	31.00	538.91	--	59.19
12/10/2004	43	--	79.59	535.32	--	58.80
12/11/2004	23	--	45.45	662.86	--	72.42

Appendix 1-B.   Measured and MOVE.1-estimated daily mean streamflow at McTier Creek near New Holland, S.C. (station 02172305) and Fishing Brook (County Line Flow) near Newcomb, N.Y. (station 0131199050).—Continued

[--, no data MOVE.1 regression technique used measured daily mean streamflow at long-term index station McTier Creek near Monetta, S.C. (station 02172300). MOVE.1 regression technique used measured daily mean streamflow at long-term index station Hudson River near Newcomb, N.Y. (station 01312000).]

Date	Daily mean streamflow, in cubic feet per second					
	McTier Creek			Fishing Brook		
	Measured for MOVE.1 long-term index station 02172300	Measured for 02172305	MOVE.1 estimate for 02172305	Measured at MOVE.1 long-term index station 01312000	Measured at 0131199050	MOVE.1 estimate for 013199050
12/12/2004	17	--	34.68	856.20	--	92.95
12/13/2004	16	--	32.84	747.42	--	81.42
12/14/2004	14	--	29.14	609.96	--	66.78
12/15/2004	14	--	29.14	490.57	--	54.01
12/16/2004	14	--	29.14	421.39	--	46.57
12/17/2004	14	--	29.14	386.67	--	42.82
12/18/2004	14	--	29.14	327.63	--	36.43
12/19/2004	14	--	29.14	308.91	--	34.40
12/20/2004	15	--	31.00	282.11	--	31.49
12/21/2004	14	--	29.14	256.27	--	28.67
12/22/2004	14	--	29.14	247.20	--	27.68
12/23/2004	17	--	34.68	255.02	--	28.54
12/24/2004	17	--	34.68	669.29	--	73.11
12/25/2004	15	--	31.00	1,060.24	--	114.49
12/26/2004	17	--	34.68	925.93	--	100.33
12/27/2004	17	--	34.68	762.04	--	82.97
12/28/2004	15	--	31.00	607.69	--	66.54
12/29/2004	14	--	29.14	518.96	--	57.05
12/30/2004	15	--	31.00	450.81	--	49.73
12/31/2004	15	--	31.00	400.50	--	44.31
1/1/2005	15	--	31.00	404.70	--	44.77
1/2/2005	14	--	29.14	464.86	--	51.24
1/3/2005	14	--	29.14	558.87	--	61.32
1/4/2005	14	--	29.14	652.48	--	71.32
1/5/2005	14	--	29.14	657.37	--	71.84
1/6/2005	13	--	27.27	585.33	--	64.15
1/7/2005	14	--	29.14	511.34	--	56.23
1/8/2005	14	--	29.14	444.82	--	49.09
1/9/2005	14	--	29.14	392.93	--	43.50
1/10/2005	14	--	29.14	353.97	--	39.29
1/11/2005	14	--	29.14	321.07	--	35.72
1/12/2005	15	--	31.00	297.38	--	33.15
1/13/2005	15	--	31.00	285.24	--	31.83
1/14/2005	50	--	91.09	389.70	--	43.15
1/15/2005	25	--	48.98	932.39	--	101.01
1/16/2005	19	--	38.31	897.04	--	97.27

Appendix 1-B.   Measured and MOVE.1-estimated daily mean streamflow at McTier Creek near New Holland, S.C. (station 02172305) and Fishing Brook (County Line Flow) near Newcomb, N.Y. (station 0131199050).—Continued

[--, no data MOVE.1 regression technique used measured daily mean streamflow at long-term index station McTier Creek near Monetta, S.C. (station 02172300). MOVE.1 regression technique used measured daily mean streamflow at long-term index station Hudson River near Newcomb, N.Y. (station 01312000).]

Date	Daily mean streamflow, in cubic feet per second					
	McTier Creek			Fishing Brook		
	Measured for MOVE.1 long-term index station 02172300	Measured for 02172305	MOVE.1 estimate for 02172305	Measured at MOVE.1 long-term index station 01312000	Measured at 0131199050	MOVE.1 estimate for 013199050
1/17/2005	17	--	34.68	694.20	--	75.76
1/18/2005	17	--	34.68	522.32	--	57.41
1/19/2005	16	--	32.84	407.76	--	45.10
1/20/2005	15	--	31.00	350.59	--	38.92
1/21/2005	15	--	31.00	311.94	--	34.73
1/22/2005	15	--	31.00	280.46	--	31.31
1/23/2005	15	--	31.00	259.66	--	29.04
1/24/2005	14	--	29.14	240.67	--	26.97
1/25/2005	14	--	29.14	228.17	--	25.60
1/26/2005	14	--	29.14	218.60	--	24.56
1/27/2005	13	--	27.27	208.15	--	23.41
1/28/2005	13	--	27.27	198.74	--	22.38
1/29/2005	15	--	31.00	186.56	--	21.04
1/30/2005	33	--	62.80	178.57	--	20.16
1/31/2005	23	--	45.45	171.32	--	19.36
2/1/2005	19	--	38.31	165.19	--	18.69
2/2/2005	17	--	34.68	159.55	--	18.06
2/3/2005	38	--	71.25	154.45	--	17.50
2/4/2005	28	--	54.21	150.72	--	17.09
2/5/2005	20	--	40.11	146.66	--	16.64
2/6/2005	18	--	36.50	143.37	--	16.28
2/7/2005	17	--	34.68	141.12	--	16.03
2/8/2005	17	--	34.68	142.33	--	16.16
2/9/2005	23	--	45.45	149.16	--	16.92
2/10/2005	26	--	50.73	165.47	--	18.72
2/11/2005	19	--	38.31	170.58	--	19.28
2/12/2005	17	--	34.68	170.49	--	19.27
2/13/2005	17	--	34.68	169.19	--	19.13
2/14/2005	17	--	34.68	163.58	--	18.51
2/15/2005	19	--	38.31	168.03	--	19.00
2/16/2005	17	--	34.68	170.90	--	19.32
2/17/2005	16	--	32.84	175.26	--	19.80
2/18/2005	15	--	31.00	175.73	--	19.85
2/19/2005	14	--	29.14	169.89	--	19.21
2/20/2005	15	--	31.00	164.89	--	18.65
2/21/2005	20	--	40.11	162.21	--	18.36

Appendix 1-B. Measured and MOVE.1-estimated daily mean streamflow at McTier Creek near New Holland, S.C. (station 02172305) and Fishing Brook (County Line Flow) near Newcomb, N.Y. (station 0131199050).—Continued

[--, no data MOVE.1 regression technique used measured daily mean streamflow at long-term index station McTier Creek near Monetta, S.C. (station 02172300). MOVE.1 regression technique used measured daily mean streamflow at long-term index station Hudson River near Newcomb, N.Y. (station 01312000).]

| Date | Daily mean streamflow, in cubic feet per second | | | | | |
| | McTier Creek | | | Fishing Brook | | |
	Measured for MOVE.1 long-term index station 02172300	Measured for 02172305	MOVE.1 estimate for 02172305	Measured at MOVE.1 long-term index station 01312000	Measured at 0131199050	MOVE.1 estimate for 013199050
2/22/2005	28	--	54.21	160.28	--	18.15
2/23/2005	22	--	43.68	157.44	--	17.83
2/24/2005	62	--	110.44	151.74	--	17.20
2/25/2005	33	--	62.80	146.96	--	16.67
2/26/2005	21	--	41.90	141.65	--	16.09
2/27/2005	21	--	41.90	136.57	--	15.52
2/28/2005	42	--	77.93	132.20	--	15.04
3/1/2005	25	--	48.98	133.17	--	15.15
3/2/2005	20	--	40.11	133.79	--	15.22
3/3/2005	18	--	36.50	130.71	--	14.87
3/4/2005	18	--	36.50	128.02	--	14.57
3/5/2005	17	--	34.68	124.75	--	14.21
3/6/2005	17	--	34.68	122.37	--	13.95
3/7/2005	16	--	32.84	121.37	--	13.84
3/8/2005	38	--	71.25	129.31	--	14.72
3/9/2005	23	--	45.45	138.86	--	15.78
3/10/2005	19	--	38.31	144.31	--	16.38
3/11/2005	17	--	34.68	144.96	--	16.45
3/12/2005	16	--	32.84	145.95	--	16.56
3/13/2005	16	--	32.84	142.88	--	16.22
3/14/2005	16	--	32.84	138.54	--	15.74
3/15/2005	16	--	32.84	134.40	--	15.28
3/16/2005	23	--	45.45	129.13	--	14.70
3/17/2005	25	--	48.98	124.31	--	14.16
3/18/2005	20	--	40.11	121.64	--	13.87
3/19/2005	18	--	36.50	118.20	--	13.48
3/20/2005	17	--	34.68	115.48	--	13.18
3/21/2005	16	--	32.84	115.40	--	13.17
3/22/2005	16	--	32.84	115.68	--	13.20
3/23/2005	23	--	45.45	117.46	--	13.40
3/24/2005	20	--	40.11	120.40	--	13.73
3/25/2005	17	--	34.68	124.45	--	14.18
3/26/2005	16	--	32.84	125.22	--	14.26
3/27/2005	106	--	178.51	128.94	--	14.68
3/28/2005	124	--	205.42	145.71	--	16.53
3/29/2005	41	--	76.27	193.45	--	21.80

Appendix 1-B. Measured and MOVE.1-estimated daily mean streamflow at McTier Creek near New Holland, S.C. (station 02172305) and Fishing Brook (County Line Flow) near Newcomb, N.Y. (station 0131199050).—Continued

[--, no data MOVE.1 regression technique used measured daily mean streamflow at long-term index station McTier Creek near Monetta, S.C. (station 02172300). MOVE.1 regression technique used measured daily mean streamflow at long-term index station Hudson River near Newcomb, N.Y. (station 01312000).]

| Date | Daily mean streamflow, in cubic feet per second | | | | | |
| | McTier Creek | | | Fishing Brook | | |
	Measured for MOVE.1 long-term index station 02172300	Measured for 02172305	MOVE.1 estimate for 02172305	Measured at MOVE.1 long-term index station 01312000	Measured at 0131199050	MOVE.1 estimate for 013199050
3/30/2005	26	--	50.73	302.65	--	33.72
3/31/2005	26	--	50.73	508.71	--	55.95
4/1/2005	32	--	61.09	999.99	--	108.14
4/2/2005	33	--	62.80	1,680.62	--	179.40
4/3/2005	24	--	47.22	2,776.03	--	292.64
4/4/2005	21	--	41.90	4,031.57	--	421.05
4/5/2005	19	--	38.31	3,028.25	--	318.54
4/6/2005	19	--	38.31	2,053.50	--	218.11
4/7/2005	28	--	54.21	1,826.61	--	194.58
4/8/2005	49	--	89.46	2,052.20	--	217.98
4/9/2005	40	--	74.60	2,093.69	--	222.27
4/10/2005	26	--	50.73	1,753.23	--	186.96
4/11/2005	21	--	41.90	1,519.54	--	162.62
4/12/2005	20	--	40.11	1,289.95	--	138.61
4/13/2005	29	--	55.94	1,065.45	--	115.04
4/14/2005	24	--	47.22	869.47	--	94.36
4/15/2005	19	--	38.31	728.37	--	79.40
4/16/2005	18	--	36.50	633.20	--	69.26
4/17/2005	17	--	34.68	581.56	--	63.75
4/18/2005	16	--	32.84	622.49	--	68.12
4/19/2005	16	--	32.84	690.99	--	75.42
4/20/2005	15	--	31.00	785.37	--	85.45
4/21/2005	14	--	29.14	1,331.00	--	142.91
4/22/2005	16	--	32.84	1,496.28	--	160.19
4/23/2005	23	--	45.45	1,324.01	--	142.18
4/24/2005	17	--	34.68	2,077.53	--	220.60
4/25/2005	15	--	31.00	3,416.91	--	358.34
4/26/2005	15	--	31.00	2,605.83	--	275.14
4/27/2005	15	--	31.00	1,791.58	--	190.94
4/28/2005	14	--	29.14	2,064.32	--	219.23
4/29/2005	14	--	29.14	2,070.53	--	219.88
4/30/2005	19	--	38.31	1,544.14	--	165.18
5/1/2005	20	--	40.11	1,326.54	--	142.45
5/2/2005	16	--	32.84	1,223.29	--	131.63
5/3/2005	14	--	29.14	1,006.90	--	108.87
5/4/2005	13	--	27.27	794.99	--	86.47

Appendix 1-B.   Measured and MOVE.1-estimated daily mean streamflow at McTier Creek near New Holland, S.C. (station 02172305) and Fishing Brook (County Line Flow) near Newcomb, N.Y. (station 0131199050).—Continued

[--, no data MOVE.1 regression technique used measured daily mean streamflow at long-term index station McTier Creek near Monetta, S.C. (station 02172300). MOVE.1 regression technique used measured daily mean streamflow at long-term index station Hudson River near Newcomb, N.Y. (station 01312000).]

| Date | Daily mean streamflow, in cubic feet per second | | | | | |
| | McTier Creek | | | Fishing Brook | | |
	Measured for MOVE.1 long-term index station 02172300	Measured for 02172305	MOVE.1 estimate for 02172305	Measured at MOVE.1 long-term index station 01312000	Measured at 0131199050	MOVE.1 estimate for 013199050
5/5/2005	14	--	29.14	644.31	--	70.45
5/6/2005	14	--	29.14	537.00	--	58.98
5/7/2005	13	--	27.27	466.57	--	51.43
5/8/2005	12	--	25.39	425.95	--	47.06
5/9/2005	11	--	23.48	395.92	--	43.82
5/10/2005	12	--	25.39	394.78	--	43.70
5/11/2005	18	--	36.50	440.58	--	48.63
5/12/2005	14	--	29.14	476.08	--	52.45
5/13/2005	12	--	25.39	409.36	--	45.27
5/14/2005	11	--	23.48	327.30	--	36.40
5/15/2005	11	--	23.48	293.51	--	32.73
5/16/2005	10	--	21.56	322.85	--	35.92
5/17/2005	10	--	21.56	319.90	--	35.59
5/18/2005	11	--	23.48	285.58	--	31.87
5/19/2005	10	--	21.56	254.36	--	28.46
5/20/2005	14	--	29.14	231.84	--	26.01
5/21/2005	18	--	36.50	216.58	--	24.34
5/22/2005	14	--	29.14	205.92	--	23.17
5/23/2005	12	--	25.39	203.75	--	22.93
5/24/2005	11	--	23.48	224.42	--	25.19
5/25/2005	10	--	21.56	261.36	--	29.23
5/26/2005	9.6	--	20.79	247.25	--	27.69
5/27/2005	8.9	--	19.43	225.29	--	25.29
5/28/2005	8.9	--	19.43	212.00	--	23.83
5/29/2005	9.2	--	20.01	203.73	--	22.93
5/30/2005	16	--	32.84	199.46	--	22.46
5/31/2005	18	--	36.50	193.49	--	21.80
6/1/2005	17	--	34.68	188.18	--	21.22
6/2/2005	155	--	250.85	179.15	--	20.22
6/3/2005	68	--	119.96	170.46	--	19.27
6/4/2005	34	--	64.50	160.96	--	18.22
6/5/2005	21	--	41.90	150.46	--	17.06
6/6/2005	16	--	32.84	140.01	--	15.90
6/7/2005	14	--	29.14	131.81	--	15.00
6/8/2005	15	--	31.00	124.96	--	14.23
6/9/2005	14	--	29.14	125.41	--	14.28

Appendix 1-B. Measured and MOVE.1-estimated daily mean streamflow at McTier Creek near New Holland, S.C. (station 02172305) and Fishing Brook (County Line Flow) near Newcomb, N.Y. (station 0131199050).—Continued

[--, no data MOVE 1 regression technique used measured daily mean streamflow at long-term index station McTier Creek near Monetta, S.C. (station 02172300). MOVE 1 regression technique used measured daily mean streamflow at long-term index station Hudson River near Newcomb, N.Y. (station 01312000).]

	Daily mean streamflow, in cubic feet per second					
	McTier Creek			Fishing Brook		
Date	Measured for MOVE.1 long-term index station 02172300	Measured for 02172305	MOVE.1 estimate for 02172305	Measured at MOVE.1 long-term index station 01312000	Measured at 0131199050	MOVE.1 estimate for 013199050
6/10/2005	14	--	29.14	163.71	--	18.52
6/11/2005	16	--	32.84	291.86	--	32.55
6/12/2005	15	--	31.00	365.31	--	40.51
6/13/2005	18	--	36.50	295.97	--	33.00
6/14/2005	15	--	31.00	301.61	--	33.61
6/15/2005	12	--	25.39	509.69	--	56.06
6/16/2005	9.5	--	20.59	624.91	--	68.38
6/17/2005	9	--	19.62	1,434.30	--	153.72
6/18/2005	8.5	--	18.64	2,219.82	--	235.32
6/19/2005	10	--	21.56	1,780.89	--	189.83
6/20/2005	9.3	--	20.21	1,209.89	--	130.22
6/21/2005	9.3	--	20.21	808.52	--	87.90
6/22/2005	9.9	--	21.37	573.55	--	62.89
6/23/2005	9	--	19.62	431.01	--	47.60
6/24/2005	8.3	--	18.25	339.69	--	37.74
6/25/2005	8.2	--	18.05	279.00	--	31.15
6/26/2005	18	--	36.50	235.63	--	26.42
6/27/2005	110	--	184.53	201.33	--	22.66
6/28/2005	119	--	197.99	175.94	--	19.87
6/29/2005	42	--	77.93	169.11	--	19.12
6/30/2005	24	--	47.22	252.35	--	28.25
7/1/2005	18	--	36.50	280.72	--	31.34
7/2/2005	15	--	31.00	249.95	--	27.98
7/3/2005	34	--	64.50	211.75	--	23.81
7/4/2005	35	--	66.19	180.00	--	20.32
7/5/2005	20	--	40.11	158.79	--	17.98
7/6/2005	16	--	32.84	150.95	--	17.11
7/7/2005	15	--	31.00	152.78	--	17.32
7/8/2005	13	--	27.27	144.27	--	16.38
7/9/2005	12	--	25.39	150.85	--	17.10
7/10/2005	13	--	27.27	246.54	--	27.61
7/11/2005	15	--	31.00	304.09	--	33.88
7/12/2005	14	--	29.14	279.99	--	31.26
7/13/2005	51	--	92.72	238.89	--	26.78
7/14/2005	61	--	108.85	228.35	--	25.62
7/15/2005	25	--	48.98	244.58	--	27.40

Appendix 1-B.   Measured and MOVE.1-estimated daily mean streamflow at McTier Creek near New Holland, S.C. (station 02172305) and Fishing Brook (County Line Flow) near Newcomb, N.Y. (station 0131199050).—Continued

[--, no data MOVE.1 regression technique used measured daily mean streamflow at long-term index station McTier Creek near Monetta, S.C. (station 02172300). MOVE.1 regression technique used measured daily mean streamflow at long-term index station Hudson River near Newcomb, N.Y. (station 01312000).]

| Date | Daily mean streamflow, in cubic feet per second | | | | | |
| | McTier Creek | | | Fishing Brook | | |
	Measured for MOVE.1 long-term index station 02172300	Measured for 02172305	MOVE.1 estimate for 02172305	Measured at MOVE.1 long-term index station 01312000	Measured at 0131199050	MOVE.1 estimate for 013199050
7/16/2005	18	--	36.50	253.71	--	28.39
7/17/2005	15	--	31.00	263.70	--	29.48
7/18/2005	15	--	31.00	266.68	--	29.81
7/19/2005	17	--	34.68	252.90	--	28.31
7/20/2005	14	--	29.14	239.07	--	26.80
7/21/2005	12	--	25.39	214.26	--	24.08
7/22/2005	11	--	23.48	187.77	--	21.17
7/23/2005	10	--	21.56	166.46	--	18.83
7/24/2005	9.9	--	21.37	147.58	--	16.74
7/25/2005	9.7	--	20.98	132.13	--	15.03
7/26/2005	10	--	21.56	120.51	--	13.74
7/27/2005	9	--	19.62	172.22	--	19.46
7/28/2005	9.1	--	19.82	510.29	--	56.12
7/29/2005	18	--	36.50	506.02	--	55.66
7/30/2005	16	--	32.84	364.29	--	40.40
7/31/2005	14	--	29.14	280.04	--	31.26
8/1/2005	14	--	29.14	229.93	--	25.80
8/2/2005	12	--	25.39	209.48	--	23.56
8/3/2005	10	--	21.56	190.81	--	21.51
8/4/2005	9.8	--	21.18	168.49	--	19.05
8/5/2005	9.1	--	19.82	149.12	--	16.91
8/6/2005	9.5	--	20.59	132.65	--	15.09
8/7/2005	11	--	23.48	118.28	--	13.49
8/8/2005	16	--	32.84	107.11	--	12.25
8/9/2005	19	--	38.31	98.25	--	11.26
8/10/2005	18	--	36.50	90.36	--	10.38
8/11/2005	20	--	40.11	83.47	--	9.60
8/12/2005	14		29.14	78.13	--	9.01
8/13/2005	33	--	62.80	75.24	--	8.68
8/14/2005	15	--	31.00	76.05	--	8.77
8/15/2005	12	--	25.39	81.53	--	9.39
8/16/2005	11	--	23.48	81.25	--	9.36
8/17/2005	12	--	25.39	76.90	--	8.87
8/18/2005	14	--	29.14	71.09	--	8.21
8/19/2005	11	--	23.48	65.97	--	7.64
8/20/2005	10	--	21.56	68.32	--	7.90

Appendix 1-B.   Measured and MOVE.1-estimated daily mean streamflow at McTier Creek near New Holland, S.C. (station 02172305) and Fishing Brook (County Line Flow) near Newcomb, N.Y. (station 0131199050).—Continued

[--, no data MOVE.1 regression technique used measured daily mean streamflow at long-term index station McTier Creek near Monetta, S.C. (station 02172300). MOVE.1 regression technique used measured daily mean streamflow at long-term index station Hudson River near Newcomb, N.Y. (station 01312000).]

Date	Daily mean streamflow, in cubic feet per second					
	McTier Creek			Fishing Brook		
	Measured for MOVE.1 long-term index station 02172300	Measured for 02172305	MOVE.1 estimate for 02172305	Measured at MOVE.1 long-term index station 01312000	Measured at 0131199050	MOVE.1 estimate for 013199050
8/21/2005	9.3	--	20.21	94.98	--	10.89
8/22/2005	35	--	66.19	110.12	--	12.58
8/23/2005	57	--	102.43	103.95	--	11.90
8/24/2005	97	--	164.88	94.09	--	10.79
8/25/2005	32	--	61.09	84.70	--	9.74
8/26/2005	17	--	34.68	77.12	--	8.89
8/27/2005	14	--	29.14	70.36	--	8.13
8/28/2005	13	--	27.27	68.83	--	7.96
8/29/2005	12	--	25.39	67.52	--	7.81
8/30/2005	12	--	25.39	66.82	--	7.73
8/31/2005	13	--	27.27	209.00	--	23.50
9/1/2005	12	--	25.39	1,305.89	--	140.28
9/2/2005	11	--	23.48	1,546.68	--	165.45
9/3/2005	10	--	21.56	1,038.96	--	112.25
9/4/2005	9.4	--	20.40	677.76	--	74.01
9/5/2005	8.9	--	19.43	486.49	--	53.57
9/6/2005	8.8	--	19.23	366.17	--	40.61
9/7/2005	9	--	19.62	290.28	--	32.38
9/8/2005	8.9	--	19.43	239.91	--	26.89
9/9/2005	8.6	--	18.84	206.90	--	23.27
9/10/2005	8.3	--	18.25	181.59	--	20.49
9/11/2005	7.9	--	17.46	157.62	--	17.85
9/12/2005	7.8	--	17.26	140.12	--	15.92
9/13/2005	7.7	--	17.06	127.20	--	14.48
9/14/2005	7.8	--	17.26	114.90	--	13.12
9/15/2005	8	--	17.66	106.12	--	12.14
9/16/2005	7.7	--	17.06	100.07	--	11.46
9/17/2005	7.6	--	16.87	100.10	--	11.47
9/18/2005	7.5	--	16.67	100.63	--	11.52
9/19/2005	7.2	--	16.07	97.94	--	11.22
9/20/2005	7.3	--	16.27	94.08	--	10.79
9/21/2005	7.5	--	16.67	91.30	--	10.48
9/22/2005	7.4	--	16.47	85.97	--	9.89
9/23/2005	7.6	--	16.87	82.01	--	9.44
9/24/2005	7.5	--	16.67	76.71	--	8.85
9/25/2005	8.5	--	18.64	73.17	--	8.45

Appendix 1-B. Measured and MOVE.1-estimated daily mean streamflow at McTier Creek near New Holland, S.C. (station 02172305) and Fishing Brook (County Line Flow) near Newcomb, N.Y. (station 0131199050).—Continued

[--, no data MOVE.1 regression technique used measured daily mean streamflow at long-term index station McTier Creek near Monetta, S.C. (station 02172300). MOVE.1 regression technique used measured daily mean streamflow at long-term index station Hudson River near Newcomb, N.Y. (station 01312000).]

| | Daily mean streamflow, in cubic feet per second | | | | | |
| | McTier Creek | | | Fishing Brook | | |
Date	Measured for MOVE.1 long-term index station 02172300	Measured for 02172305	MOVE.1 estimate for 02172305	Measured at MOVE.1 long-term index station 01312000	Measured at 0131199050	MOVE.1 estimate for 013199050
9/26/2005	8	--	17.66	82.53	--	9.50
9/27/2005	8.8	--	19.23	320.09	--	35.62
9/28/2005	8.6	--	18.84	442.78	--	48.87
9/29/2005	8.7	--	19.03	329.82	--	36.67
9/30/2005	8.8	--	19.23	384.82	--	42.62
10/1/2005	8.6	--	18.84	359.19	--	39.85
10/2/2005	8	--	17.66	282.31	--	31.51
10/3/2005	7.6	--	16.87	230.42	--	25.85
10/4/2005	7.7	--	17.06	195.20	--	21.99
10/5/2005	7.8	--	17.26	170.31	--	19.25
10/6/2005	22	--	43.68	151.01	--	17.12
10/7/2005	34	--	64.50	138.73	--	15.76
10/8/2005	76	--	132.52	438.06	--	48.36
10/9/2005	23	--	45.45	1,165.06	--	125.51
10/10/2005	46	--	84.54	998.32	--	107.97
10/11/2005	21	--	41.90	690.68	--	75.39
10/12/2005	16	--	32.84	513.93	--	56.51
10/13/2005	14	--	29.14	412.19	--	45.57
10/14/2005	13	--	27.27	380.66	--	42.17
10/15/2005	12	--	25.39	934.36	--	101.22
10/16/2005	11	--	23.48	1,820.08	--	193.90
10/17/2005	11	--	23.48	2,352.09	--	248.98
10/18/2005	11	--	23.48	2,383.53	--	252.23
10/19/2005	10	--	21.56	2,011.85	--	213.80
10/20/2005	10	--	21.56	1,780.16	--	189.76
10/21/2005	11	--	23.48	1,328.35	--	142.64
10/22/2005	11	--	23.48	986.71	--	106.74
10/23/2005	10	--	21.56	877.74	--	95.23
10/24/2005	10	--	21.56	991.15	--	107.21
10/25/2005	10	--	21.56	1,014.89	--	109.71
10/26/2005	10	--	21.56	1,128.47	--	121.67
10/27/2005	10	--	21.56	1,055.01	--	113.94
10/28/2005	10	--	21.56	878.27	--	95.29
10/29/2005	10	--	21.56	713.56	--	77.82
10/30/2005	11	--	23.48	597.62	--	65.47
10/31/2005	11	--	23.48	545.18	--	59.86

Appendix 1-B.   Measured and MOVE.1-estimated daily mean streamflow at McTier Creek near New Holland, S.C. (station 02172305) and Fishing Brook (County Line Flow) near Newcomb, N.Y. (station 0131199050).—Continued

[--, no data MOVE.1 regression technique used measured daily mean streamflow at long-term index station McTier Creek near Monetta, S.C. (station 02172300). MOVE.1 regression technique used measured daily mean streamflow at long-term index station Hudson River near Newcomb, N.Y. (station 01312000).]

| Date | Daily mean streamflow, in cubic feet per second | | | | | |
| | McTier Creek | | | Fishing Brook | | |
	Measured for MOVE.1 long-term index station 02172300	Measured for 02172305	MOVE.1 estimate for 02172305	Measured at MOVE.1 long-term index station 01312000	Measured at 0131199050	MOVE.1 estimate for 013199050
11/1/2005	11	--	23.48	572.28	--	62.76
11/2/2005	10	--	21.56	648.65	--	70.91
11/3/2005	10	--	21.56	622.22	--	68.09
11/4/2005	10	--	21.56	528.89	--	58.11
11/5/2005	11	--	23.48	483.00	--	53.19
11/6/2005	11	--	23.48	475.88	--	52.43
11/7/2005	11	--	23.48	610.65	--	66.86
11/8/2005	11	--	23.48	723.76	--	78.91
11/9/2005	11	--	23.48	676.64	--	73.89
11/10/2005	11	--	23.48	1,285.85	--	138.18
11/11/2005	11	--	23.48	1,711.40	--	182.61
11/12/2005	11	--	23.48	1,388.26	--	148.91
11/13/2005	11	--	23.48	1,078.20	--	116.38
11/14/2005	11	--	23.48	858.94	--	93.24
11/15/2005	11	--	23.48	752.46	--	81.95
11/16/2005	12	--	25.39	1,163.46	--	125.34
11/17/2005	13	--	27.27	1,908.79	--	203.11
11/18/2005	13	--	27.27	1,908.88	--	203.12
11/19/2005	12	--	25.39	1,425.68	--	152.82
11/20/2005	12	--	25.39	1,095.14	--	118.16
11/21/2005	49	--	89.46	860.42	--	93.40
11/22/2005	33	--	62.80	713.64	--	77.83
11/23/2005	18	--	36.50	624.31	--	68.31
11/24/2005	15	--	31.00	519.17	--	57.07
11/25/2005	13	--	27.27	456.76	--	50.37
11/26/2005	13	--	27.27	415.13	--	45.89
11/27/2005	13	--	27.27	374.38	--	41.49
11/28/2005	14	--	29.14	359.55	--	39.89
11/29/2005	18	--	36.50	399.03	--	44.16
11/30/2005	16	--	32.84	1,294.45	--	139.09
12/1/2005	15	--	31.00	2,278.37	--	241.37
12/2/2005	14	--	29.14	1,815.57	--	193.44
12/3/2005	14	--	29.14	1,320.60	--	141.82
12/4/2005	14	--	29.14	1,011.03	--	109.31
12/5/2005	31	--	59.38	785.09	--	85.42
12/6/2005	37	--	69.57	628.71	--	68.78

Appendix 1-B.   Measured and MOVE.1-estimated daily mean streamflow at McTier Creek near New Holland, S.C. (station 02172305) and Fishing Brook (County Line Flow) near Newcomb, N.Y. (station 0131199050).—Continued

[--, no data MOVE.1 regression technique used measured daily mean streamflow at long-term index station McTier Creek near Monetta, S.C. (station 02172300). MOVE.1 regression technique used measured daily mean streamflow at long-term index station Hudson River near Newcomb, N.Y. (station 01312000).]

| Date | Daily mean streamflow, in cubic feet per second | | | | | |
| | McTier Creek | | | Fishing Brook | | |
	Measured for MOVE.1 long-term index station 02172300	Measured for 02172305	MOVE.1 estimate for 02172305	Measured at MOVE.1 long-term index station 01312000	Measured at 0131199050	MOVE.1 estimate for 013199050
12/7/2005	21	--	41.90	517.20	--	56.86
12/8/2005	18	--	36.50	427.05	--	47.18
12/9/2005	28	--	54.21	389.13	--	43.09
12/10/2005	21	--	41.90	374.24	--	41.48
12/11/2005	18	--	36.50	346.92	--	38.52
12/12/2005	17	--	34.68	330.15	--	36.71
12/13/2005	16	--	32.84	290.96	--	32.45
12/14/2005	15	--	31.00	262.54	--	29.36
12/15/2005	29	--	55.94	246.70	--	27.63
12/16/2005	34	--	64.50	254.68	--	28.50
12/17/2005	22	--	43.68	257.89	--	28.85
12/18/2005	33	--	62.80	254.47	--	28.48
12/19/2005	24	--	47.22	246.27	--	27.58
12/20/2005	20	--	40.11	236.94	--	26.56
12/21/2005	18	--	36.50	225.45	--	25.31
12/22/2005	17	--	34.68	217.65	--	24.45
12/23/2005	17	--	34.68	213.82	--	24.03
12/24/2005	16	--	32.84	210.36	--	23.65
12/25/2005	37	--	69.57	212.24	--	23.86
12/26/2005	27	--	52.47	294.13	--	32.80
12/27/2005	20	--	40.11	422.50	--	46.69
12/28/2005	20	--	40.11	424.84	--	46.94
12/29/2005	34	--	64.50	388.08	--	42.97
12/30/2005	21	--	41.90	421.44	--	46.57
12/31/2005	18	--	36.50	441.37	--	48.72
1/1/2006	17	--	34.68	426.06	--	47.07
1/2/2006	58	--	104.04	388.66	--	43.04
1/3/2006	58	--	104.04	354.56	--	39.35
1/4/2006	29	--	55.94	320.72	--	35.68
1/5/2006	24	--	47.22	311.62	--	34.70
1/6/2006	21	--	41.90	296.36	--	33.04
1/7/2006	20	--	40.11	268.94	--	30.05
1/8/2006	19	--	38.31	257.06	--	28.76
1/9/2006	18	--	36.50	248.17	--	27.79
1/10/2006	18	--	36.50	237.99	--	26.68
1/11/2006	18	--	36.50	225.35	--	25.29

Appendix 1-B. Measured and MOVE.1-estimated daily mean streamflow at McTier Creek near New Holland, S.C. (station 02172305) and Fishing Brook (County Line Flow) near Newcomb, N.Y. (station 0131199050).—Continued

[--, no data MOVE 1 regression technique used measured daily mean streamflow at long-term index station McTier Creek near Monetta, S.C. (station 02172300). MOVE 1 regression technique used measured daily mean streamflow at long-term index station Hudson River near Newcomb, N.Y. (station 01312000).]

	Daily mean streamflow, in cubic feet per second					
	McTier Creek			Fishing Brook		
Date	Measured for MOVE.1 long-term index station 02172300	Measured for 02172305	MOVE.1 estimate for 02172305	Measured at MOVE.1 long-term index station 01312000	Measured at 0131199050	MOVE.1 estimate for 013199050
1/12/2006	17	--	34.68	250.51	--	28.04
1/13/2006	19	--	38.31	307.89	--	34.29
1/14/2006	27	--	52.47	351.18	--	38.98
1/15/2006	20	--	40.11	548.13	--	60.17
1/16/2006	18	--	36.50	653.60	--	71.44
1/17/2006	17	--	34.68	529.05	--	58.13
1/18/2006	38	--	71.25	639.86	--	69.97
1/19/2006	24	--	47.22	1,494.27	--	159.98
1/20/2006	20	--	40.11	1,813.63	--	193.23
1/21/2006	19	--	38.31	1,529.53	--	163.66
1/22/2006	22	--	43.68	1,298.63	--	139.52
1/23/2006	23	--	45.45	1,145.79	--	123.49
1/24/2006	24	--	47.22	963.37	--	104.28
1/25/2006	21	--	41.90	789.16	--	85.85
1/26/2006	18	--	36.50	650.20	--	71.08
1/27/2006	17	--	34.68	518.24	--	56.97
1/28/2006	17	--	34.68	461.37	--	50.87
1/29/2006	18	--	36.50	427.54	--	47.23
1/30/2006	19	--	38.31	402.57	--	44.54
1/31/2006	17	--	34.68	420.91	--	46.51
2/1/2006	16	--	32.84	461.38	--	50.87
2/2/2006	16	--	32.84	449.23	--	49.56
2/3/2006	16	--	32.84	447.08	--	49.33
2/4/2006	17	--	34.68	590.78	--	64.74
2/5/2006	16	--	32.84	853.20	--	92.63
2/6/2006	16	--	32.84	1,207.39	--	129.96
2/7/2006	21	--	41.90	1,144.46	--	123.35
2/8/2006	17	--	34.68	924.96	--	100.22
2/9/2006	16	--	32.84	704.31	--	76.84
2/10/2006	16	--	32.84	582.70	--	63.87
2/11/2006	18	--	36.50	482.67	--	53.16
2/12/2006	19	--	38.31	426.36	--	47.10
2/13/2006	16	--	32.84	384.13	--	42.55
2/14/2006	15	--	31.00	351.97	--	39.07
2/15/2006	15	--	31.00	329.98	--	36.69
2/16/2006	14	--	29.14	313.81	--	34.93

Appendix 1-B. Measured and MOVE.1-estimated daily mean streamflow at McTier Creek near New Holland, S.C. (station 02172305) and Fishing Brook (County Line Flow) near Newcomb, N.Y. (station 0131199050).—Continued

[--, no data MOVE.1 regression technique used measured daily mean streamflow at long-term index station McTier Creek near Monetta, S.C. (station 02172300). MOVE.1 regression technique used measured daily mean streamflow at long-term index station Hudson River near Newcomb, N.Y. (station 01312000).]

	Daily mean streamflow, in cubic feet per second					
	McTier Creek			Fishing Brook		
Date	Measured for MOVE.1 long-term index station 02172300	Measured for 02172305	MOVE.1 estimate for 02172305	Measured at MOVE.1 long-term index station 01312000	Measured at 0131199050	MOVE.1 estimate for 013199050
2/17/2006	15	--	31.00	324.74	--	36.12
2/18/2006	15	--	31.00	388.77	--	43.05
2/19/2006	15	--	31.00	407.35	--	45.05
2/20/2006	16	--	32.84	383.41	--	42.47
2/21/2006	16	--	32.84	348.86	--	38.73
2/22/2006	17	--	34.68	319.98	--	35.60
2/23/2006	45	--	82.89	297.49	--	33.16
2/24/2006	24	--	47.22	278.27	--	31.07
2/25/2006	25	--	48.98	260.92	--	29.18
2/26/2006	59	--	105.64	252.43	--	28.25
2/27/2006	27	--	52.47	240.00	--	26.90
2/28/2006	21	--	41.90	230.00	--	25.80
3/1/2006	19	--	38.31	220.00	--	24.71
3/2/2006	18	--	36.50	210.00	--	23.61
3/3/2006	17	--	34.68	205.55	--	23.13
3/4/2006	16	--	32.84	202.22	--	22.76
3/5/2006	15	--	31.00	191.45	--	21.58
3/6/2006	16	--	32.84	180.66	--	20.39
3/7/2006	14	--	29.14	174.69	--	19.73
3/8/2006	14	--	29.14	169.13	--	19.12
3/9/2006	14	--	29.14	167.13	--	18.90
3/10/2006	14	--	29.14	178.03	--	20.10
3/11/2006	15	--	31.00	214.41	--	24.10
3/12/2006	15	--	31.00	269.83	--	30.15
3/13/2006	15	--	31.00	333.04	--	37.02
3/14/2006	14	--	29.14	736.29	--	80.24
3/15/2006	13	--	27.27	1,388.26	--	148.90
3/16/2006	13	--	27.27	1,338.05	--	143.65
3/17/2006	13	--	27.27	1,039.35	--	112.29
3/18/2006	13	--	27.27	772.66	--	84.10
3/19/2006	13	--	27.27	605.93	--	66.35
3/20/2006	13	--	27.27	492.06	--	54.16
3/21/2006	38	--	71.25	412.65	--	45.62
3/22/2006	23	--	45.45	368.04	--	40.81
3/23/2006	17	--	34.68	327.77	--	36.45
3/24/2006	15	--	31.00	302.72	--	33.73

Appendix 1-B.   Measured and MOVE.1-estimated daily mean streamflow at McTier Creek near New Holland, S.C. (station 02172305) and Fishing Brook (County Line Flow) near Newcomb, N.Y. (station 0131199050).—Continued

[--, no data MOVE.1 regression technique used measured daily mean streamflow at long-term index station McTier Creek near Monetta, S.C. (station 02172300). MOVE.1 regression technique used measured daily mean streamflow at long-term index station Hudson River near Newcomb, N.Y. (station 01312000).]

| | Daily mean streamflow, in cubic feet per second | | | | | |
| | McTier Creek | | | Fishing Brook | | |
Date	Measured for MOVE.1 long-term index station 02172300	Measured for 02172305	MOVE.1 estimate for 02172305	Measured at MOVE.1 long-term index station 01312000	Measured at 0131199050	MOVE.1 estimate for 013199050
3/25/2006	15	--	31.00	286.56	--	31.97
3/26/2006	14	--	29.14	281.68	--	31.44
3/27/2006	14	--	29.14	273.72	--	30.58
3/28/2006	14	--	29.14	285.82	--	31.89
3/29/2006	14	--	29.14	318.09	--	35.40
3/30/2006	14	--	29.14	370.00	--	41.02
3/31/2006	13	--	27.27	520.00	--	57.16
4/1/2006	13	--	27.27	797.58	--	86.74
4/2/2006	13	--	27.27	1,461.75	--	156.58
4/3/2006	13	--	27.27	1,514.86	--	162.13
4/4/2006	12	--	25.39	1,394.68	--	149.58
4/5/2006	12	--	25.39	1,335.58	--	143.39
4/6/2006	12	--	25.39	1,112.56	--	120.00
4/7/2006	12	--	25.39	909.04	--	98.54
4/8/2006	14	--	29.14	816.70	--	88.77
4/9/2006	15	--	31.00	743.13	--	80.96
4/10/2006	14	--	29.14	663.30	--	72.47
4/11/2006	12	--	25.39	614.23	--	67.24
4/12/2006	12	--	25.39	600.23	--	65.74
4/13/2006	12	--	25.39	747.25	--	81.40
4/14/2006	12	--	25.39	1,039.81	--	112.34
4/15/2006	11	--	23.48	1,071.82	--	115.71
4/16/2006	10	--	21.56	1,171.49	--	126.19
4/17/2006	9.7	--	20.98	1,055.92	--	114.04
4/18/2006	9.1	--	19.82	846.55	--	91.93
4/19/2006	15	--	31.00	730.40	--	79.61
4/20/2006	17	--	34.68	747.14	--	81.39
4/21/2006	13	--	27.27	735.93	--	80.20
4/22/2006	12	--	25.39	687.02	--	75.00
4/23/2006	12	--	25.39	927.05	--	100.44
4/24/2006	11	--	23.48	1,787.72	--	190.54
4/25/2006	9.3	--	20.21	1,915.19	--	203.78
4/26/2006	9.2	--	20.01	1,573.71	--	168.27
4/27/2006	12	--	25.39	1,233.49	--	132.70
4/28/2006	12	--	25.39	975.17	--	105.52
4/29/2006	10	--	21.56	758.56	--	82.60

Appendix 1-B.   Measured and MOVE.1-estimated daily mean streamflow at McTier Creek near New Holland, S.C. (station 02172305) and Fishing Brook (County Line Flow) near Newcomb, N.Y. (station 0131199050).—Continued

[--, no data MOVE.1 regression technique used measured daily mean streamflow at long-term index station McTier Creek near Monetta, S.C. (station 02172300). MOVE.1 regression technique used measured daily mean streamflow at long-term index station Hudson River near Newcomb, N.Y. (station 01312000).]

| | Daily mean streamflow, in cubic feet per second | | | | | |
| | McTier Creek | | | Fishing Brook | | |
Date	Measured for MOVE.1 long-term index station 02172300	Measured for 02172305	MOVE.1 estimate for 02172305	Measured at MOVE.1 long-term index station 01312000	Measured at 0131199050	MOVE.1 estimate for 013199050
4/30/2006	10	--	21.56	611.61	--	66.96
5/1/2006	11	--	23.48	525.79	--	57.78
5/2/2006	9.8	--	21.18	502.79	--	55.32
5/3/2006	8.1	--	17.86	513.09	--	56.42
5/4/2006	7.6	--	16.87	516.97	--	56.84
5/5/2006	7.9	--	17.46	528.86	--	58.11
5/6/2006	8.9	--	19.43	523.52	--	57.54
5/7/2006	11	--	23.48	474.36	--	52.26
5/8/2006	14	--	29.14	413.34	--	45.70
5/9/2006	12	--	25.39	360.83	--	40.03
5/10/2006	11	--	23.48	338.41	--	37.60
5/11/2006	10	--	21.56	330.69	--	36.77
5/12/2006	9.1	--	19.82	360.66	--	40.01
5/13/2006	8.6	--	18.84	664.28	--	72.58
5/14/2006	15	--	31.00	1,033.12	--	111.63
5/15/2006	16	--	32.84	1,163.32	--	125.33
5/16/2006	12	--	25.39	1,132.93	--	122.14
5/17/2006	10	--	21.56	1,161.74	--	125.16
5/18/2006	9.2	--	20.01	1,451.86	--	155.55
5/19/2006	9.5	--	20.59	1,721.11	--	183.62
5/20/2006	11	--	23.48	1,731.25	--	184.67
5/21/2006	13	--	27.27	1,619.34	--	173.02
5/22/2006	11	--	23.48	1,400.34	--	150.17
5/23/2006	9.3	--	20.21	1,195.38	--	128.70
5/24/2006	8.3	--	18.25	1,013.36	--	109.55
5/25/2006	7.6	--	16.87	848.74	--	92.16
5/26/2006	7.3	--	16.27	710.59	--	77.50
5/27/2006	6.8	--	15.27	634.38	--	69.39
5/28/2006	6.2	--	14.05	575.03	--	63.05
5/29/2006	5.8	--	13.24	499.41	--	54.95
5/30/2006	5.8	--	13.24	458.71	--	50.58
5/31/2006	5.5	--	12.63	566.11	--	62.10
6/1/2006	5.2	--	12.01	781.67	--	85.05
6/2/2006	10	--	21.56	1,044.72	--	112.86
6/3/2006	34	--	64.50	1,037.88	--	112.14
6/4/2006	14	--	29.14	1,220.52	--	131.34

Appendix 1-B.   Measured and MOVE.1-estimated daily mean streamflow at McTier Creek near New Holland, S.C. (station 02172305) and Fishing Brook (County Line Flow) near Newcomb, N.Y. (station 0131199050).—Continued

[--, no data MOVE 1 regression technique used measured daily mean streamflow at long-term index station McTier Creek near Monetta, S.C. (station 02172300). MOVE 1 regression technique used measured daily mean streamflow at long-term index station Hudson River near Newcomb, N.Y. (station 01312000).]

Date	Daily mean streamflow, in cubic feet per second					
	McTier Creek			Fishing Brook		
	Measured for MOVE.1 long-term index station 02172300	Measured for 02172305	MOVE.1 estimate for 02172305	Measured at MOVE.1 long-term index station 01312000	Measured at 0131199050	MOVE.1 estimate for 013199050
6/5/2006	9.5	--	20.59	1,221.70	--	131.46
6/6/2006	8.1	--	17.86	1,035.36	--	111.87
6/7/2006	7.4	--	16.47	816.85	--	88.78
6/8/2006	6.7	--	15.07	665.61	--	72.72
6/9/2006	6	--	13.65	569.04	--	62.41
6/10/2006	5.8	--	13.24	518.98	--	57.05
6/11/2006	5.3	--	12.21	493.57	--	54.33
6/12/2006	4.9	--	11.38	475.72	--	52.41
6/13/2006	7.1	--	15.87	441.08	--	48.69
6/14/2006	117	--	195.01	398.92	--	44.14
6/15/2006	29	--	55.94	383.58	--	42.49
6/16/2006	13	--	27.27	337.43	--	37.50
6/17/2006	9.3	--	20.21	298.81	--	33.30
6/18/2006	8	--	17.66	275.02	--	30.72
6/19/2006	7.4	--	16.47	261.30	--	29.22
6/20/2006	7.2	--	16.07	304.10	--	33.88
6/21/2006	6.7	--	15.07	308.90	--	34.40
6/22/2006	6.4	--	14.46	278.26	--	31.07
6/23/2006	6.2	--	14.05	256.15	--	28.66
6/24/2006	6	--	13.65	233.62	--	26.20
6/25/2006	6.4	--	14.46	208.52	--	23.45
6/26/2006	9.5	--	20.59	227.06	--	25.48
6/27/2006	84	--	144.95	509.24	--	56.01
6/28/2006	10	--	21.56	1,611.93	--	172.25
6/29/2006	8	--	17.66	3,058.65	--	321.66
6/30/2006	7	--	15.67	2,621.85	--	276.79
7/1/2006	6.5	--	14.66	1,726.93	--	184.22
7/2/2006	6.3	--	14.26	1,282.17	--	137.80
7/3/2006	6.1	--	13.85	1,078.56	--	116.42
7/4/2006	5.6	--	12.83	824.06	--	89.55
7/5/2006	4.9	--	11.38	609.46	--	66.73
7/6/2006	5.7	--	13.04	467.82	--	51.56
7/7/2006	6.9	--	15.47	374.87	--	41.55
7/8/2006	6.7	--	15.07	313.02	--	34.85
7/9/2006	6	--	13.65	268.02	--	29.95
7/10/2006	5.7	--	13.04	234.23	--	26.27

Appendix 1-B.   Measured and MOVE.1-estimated daily mean streamflow at McTier Creek near New Holland, S.C. (station 02172305) and Fishing Brook (County Line Flow) near Newcomb, N.Y. (station 0131199050).—Continued

[--, no data MOVE.1 regression technique used measured daily mean streamflow at long-term index station McTier Creek near Monetta, S.C. (station 02172300). MOVE.1 regression technique used measured daily mean streamflow at long-term index station Hudson River near Newcomb, N.Y. (station 01312000).]

	Daily mean streamflow, in cubic feet per second					
	McTier Creek			Fishing Brook		
Date	Measured for MOVE.1 long-term index station 02172300	Measured for 02172305	MOVE.1 estimate for 02172305	Measured at MOVE.1 long-term index station 01312000	Measured at 0131199050	MOVE.1 estimate for 013199050
7/11/2006	5.4	--	12.42	206.92	--	23.28
7/12/2006	5.1	--	11.80	214.70	--	24.13
7/13/2006	4.9	--	11.38	564.19	--	61.89
7/14/2006	5.2	--	12.01	788.12	--	85.74
7/15/2006	5	--	11.59	620.69	--	67.93
7/16/2006	4.9	--	11.38	486.25	--	53.54
7/17/2006	4.5	--	10.55	394.10	--	43.62
7/18/2006	4.1	--	9.71	320.34	--	35.64
7/19/2006	4	--	9.49	267.11	--	29.86
7/20/2006	3.9	--	9.28	226.84	--	25.46
7/21/2006	3.8	--	9.07	199.78	--	22.49
7/22/2006	3.4	--	8.21	210.02	--	23.62
7/23/2006	4	--	9.49	338.03	--	37.56
7/24/2006	6.4	--	14.46	343.18	--	38.12
7/25/2006	14	--	29.14	297.14	--	33.12
7/26/2006	9.1	--	19.82	288.34	--	32.17
7/27/2006	6.5	--	14.66	270.98	--	30.28
7/28/2006	5.6	--	12.83	334.37	--	37.16
7/29/2006	6.5	--	14.66	808.34	--	87.88
7/30/2006	7.3	--	16.27	1,091.16	--	117.74
7/31/2006	6.5	--	14.66	917.53	--	99.44
8/1/2006	5.6	--	12.83	949.07	--	102.77
8/2/2006	5.2	--	12.01	1,440.87	--	154.40
8/3/2006	4.7	--	10.97	1,105.42	--	119.24
8/4/2006	4.3	--	10.13	781.51	--	85.04
8/5/2006	5.6	--	12.83	573.46	--	62.88
8/6/2006	7.7	--	17.06	433.92	--	47.91
8/7/2006	7.3	--	16.27	348.98	--	38.75
8/8/2006	5.8	--	13.24	295.01	--	32.89
8/9/2006	5.8	--	13.24	251.97	--	28.20
8/10/2006	5.5	--	12.63	218.53	--	24.55
8/11/2006	5.3	--	12.21	199.33	--	22.44
8/12/2006	8.2	--	18.05	181.32	--	20.46
8/13/2006	9.2	--	20.01	160.91	--	18.21
8/14/2006	7.5	--	16.67	145.22	--	16.48
8/15/2006	6.4	--	14.46	136.64	--	15.53

Appendix 1-B.   Measured and MOVE.1-estimated daily mean streamflow at McTier Creek near New Holland, S.C. (station 02172305) and Fishing Brook (County Line Flow) near Newcomb, N.Y. (station 0131199050).—Continued

[--, no data MOVE.1 regression technique used measured daily mean streamflow at long-term index station McTier Creek near Monetta, S.C. (station 02172300). MOVE.1 regression technique used measured daily mean streamflow at long-term index station Hudson River near Newcomb, N.Y. (station 01312000).]

| Date | Daily mean streamflow, in cubic feet per second | | | | | |
| | McTier Creek | | | Fishing Brook | | |
	Measured for MOVE.1 long-term index station 02172300	Measured for 02172305	MOVE.1 estimate for 02172305	Measured at MOVE.1 long-term index station 01312000	Measured at 0131199050	MOVE.1 estimate for 013199050
8/16/2006	15	--	31.00	128.84	--	14.67
8/17/2006	16	--	32.84	119.58	--	13.64
8/18/2006	8.1	--	17.86	110.37	--	12.61
8/19/2006	6.4	--	14.46	104.96	--	12.01
8/20/2006	5.9	--	13.44	121.26	--	13.82
8/21/2006	6	--	13.65	129.79	--	14.77
8/22/2006	5.1	--	11.80	122.39	--	13.95
8/23/2006	5.7	--	13.04	115.94	--	13.23
8/24/2006	9.3	--	20.21	111.30	--	12.71
8/25/2006	8.6	--	18.84	105.12	--	12.03
8/26/2006	6.9	--	15.47	97.63	--	11.19
8/27/2006	5.8	--	13.24	100.87	--	11.55
8/28/2006	5.5	--	12.63	126.13	--	14.36
8/29/2006	5.2	--	12.01	141.38	--	16.06
8/30/2006	6.8	--	15.27	139.34	--	15.83
8/31/2006	8.7	--	19.03	126.97	--	14.46
9/1/2006	7.9	--	17.46	113.60	--	12.97
9/2/2006	6.6	--	14.86	101.16	--	11.58
9/3/2006	6.1	--	13.85	95.48	--	10.95
9/4/2006	6.2	--	14.05	100.88	--	11.55
9/5/2006	8.2	--	18.05	109.26	--	12.49
9/6/2006	15	--	31.00	110.44	--	12.62
9/7/2006	9	--	19.62	113.40	--	12.95
9/8/2006	7.7	--	17.06	112.27	--	12.82
9/9/2006	6.8	--	15.27	106.52	--	12.18
9/10/2006	6.6	--	14.86	104.51	--	11.96
9/11/2006	6.8	--	15.27	97.41	--	11.17
9/12/2006	6.6	--	14.86	89.43	--	10.27
9/13/2006	31	--	59.38	87.88	--	10.10
9/14/2006	46	--	84.54	101.12	--	11.58
9/15/2006	14	--	29.14	123.91	--	14.12
9/16/2006	9.7	--	20.98	144.16	--	16.36
9/17/2006	8.6	--	18.84	142.35	--	16.16
9/18/2006	7.7	--	17.06	129.50	--	14.74
9/19/2006	7.7	--	17.06	117.69	--	13.43
9/20/2006	7	--	15.67	111.42	--	12.73

Appendix 1-B.   Measured and MOVE.1-estimated daily mean streamflow at McTier Creek near New Holland, S.C. (station 02172305) and Fishing Brook (County Line Flow) near Newcomb, N.Y. (station 0131199050).—Continued

[--, no data MOVE.1 regression technique used measured daily mean streamflow at long-term index station McTier Creek near Monetta, S.C. (station 02172300). MOVE.1 regression technique used measured daily mean streamflow at long-term index station Hudson River near Newcomb, N.Y. (station 01312000).]

| Date | Daily mean streamflow, in cubic feet per second | | | | | |
| | McTier Creek | | | Fishing Brook | | |
	Measured for MOVE.1 long-term index station 02172300	Measured for 02172305	MOVE.1 estimate for 02172305	Measured at MOVE.1 long-term index station 01312000	Measured at 0131199050	MOVE.1 estimate for 013199050
9/21/2006	6.5	--	14.66	105.36	--	12.05
9/22/2006	6.6	--	14.86	100.76	--	11.54
9/23/2006	6.6	--	14.86	99.78	--	11.43
9/24/2006	6.8	--	15.27	104.10	--	11.91
9/25/2006	6.1	--	13.85	114.31	--	13.05
9/26/2006	5.9	--	13.44	118.22	--	13.49
9/27/2006	6	--	13.65	114.41	--	13.06
9/28/2006	5.9	--	13.44	106.94	--	12.23
9/29/2006	5.6	--	12.83	122.74	--	13.99
9/30/2006	6.1	--	13.85	188.56	--	21.26
10/1/2006	5.5	--	12.63	249.58	--	27.94
10/2/2006	5.3	--	12.21	632.65	--	69.20
10/3/2006	5.5	--	12.63	613.64	--	67.18
10/4/2006	5.5	--	12.63	428.28	--	47.31
10/5/2006	5.6	--	12.83	345.36	--	38.35
10/6/2006	5.4	--	12.42	328.26	--	36.50
10/7/2006	5	--	11.59	284.68	--	31.77
10/8/2006	6.9	--	15.47	245.12	--	27.46
10/9/2006	9.3	--	20.21	211.39	--	23.77
10/10/2006	9.4	--	20.40	187.68	--	21.16
10/11/2006	8.2	--	18.05	168.03	--	19.00
10/12/2006	7	--	15.67	158.49	--	17.95
10/13/2006	5.9	--	13.44	157.40	--	17.83
10/14/2006	6.5	--	14.66	154.98	--	17.56
10/15/2006	5.9	--	13.44	146.68	--	16.64
10/16/2006	6.6	--	14.86	140.17	--	15.92
10/17/2006	7.8	--	17.26	139.82	--	15.88
10/18/2006	8.2	--	18.05	319.94	--	35.60
10/19/2006	7.9	--	17.46	848.59	--	92.15
10/20/2006	7.4	--	16.47	1,163.07	--	125.30
10/21/2006	6.9	--	15.47	2,118.23	--	224.81
10/22/2006	8.5	--	18.64	2,071.36	--	219.96
10/23/2006	9.1	--	19.82	1,612.76	--	172.34
10/24/2006	8.4	--	18.45	1,413.78	--	151.57
10/25/2006	7.9	--	17.46	1,200.59	--	129.24
10/26/2006	8.1	--	17.86	1,012.02	--	109.41

Appendix 1-B.   Measured and MOVE.1-estimated daily mean streamflow at McTier Creek near New Holland, S.C. (station 02172305) and Fishing Brook (County Line Flow) near Newcomb, N.Y. (station 0131199050).—Continued

[--, no data MOVE.1 regression technique used measured daily mean streamflow at long-term index station McTier Creek near Monetta, S.C. (station 02172300). MOVE.1 regression technique used measured daily mean streamflow at long-term index station Hudson River near Newcomb, N.Y. (station 01312000).]

| Date | Daily mean streamflow, in cubic feet per second | | | | | |
| | McTier Creek | | | Fishing Brook | | |
	Measured for MOVE.1 long-term index station 02172300	Measured for 02172305	MOVE.1 estimate for 02172305	Measured at MOVE.1 long-term index station 01312000	Measured at 0131199050	MOVE.1 estimate for 013199050
10/27/2006	12	--	25.39	816.39	--	88.74
10/28/2006	29	--	55.94	878.12	--	95.27
10/29/2006	13	--	27.27	2,047.80	--	217.52
10/30/2006	9.7	--	20.98	2,318.56	--	245.52
10/31/2006	8.7	--	19.03	1,727.89	--	184.32
11/1/2006	8.6	--	18.84	1,379.85	--	148.03
11/2/2006	8.5	--	18.64	1,276.55	--	137.21
11/3/2006	8.2	--	18.05	1,079.66	--	116.53
11/4/2006	8.2	--	18.05	858.03	--	93.15
11/5/2006	8.3	--	18.25	683.09	--	74.58
11/6/2006	8.4	--	18.45	577.90	--	63.36
11/7/2006	9.2	--	20.01	484.79	--	53.38
11/8/2006	11	--	23.48	426.05	--	47.07
11/9/2006	10	--	21.56	437.50	--	48.30
11/10/2006	9.4	--	20.40	509.05	--	55.99
11/11/2006	9.2	--	20.01	491.97	--	54.16
11/12/2006	9.8	--	21.18	813.63	--	88.44
11/13/2006	9.7	--	20.98	1,370.28	--	147.02
11/14/2006	9.8	--	21.18	1,233.07	--	132.65
11/15/2006	10	--	21.56	1,115.17	--	120.27
11/16/2006	67	--	118.38	1,022.76	--	110.54
11/17/2006	23	--	45.45	1,984.91	--	211.01
11/18/2006	14	--	29.14	2,759.48	--	290.94
11/19/2006	12	--	25.39	2,047.11	--	217.45
11/20/2006	10	--	21.56	1,426.95	--	152.95
11/21/2006	13	--	27.27	1,071.92	--	115.72
11/22/2006	52	--	94.35	822.11	--	89.34
11/23/2006	36	--	67.88	657.27	--	71.83
11/24/2006	19	--	38.31	549.80	--	60.35
11/25/2006	16	--	32.84	471.66	--	51.97
11/26/2006	13	--	27.27	414.40	--	45.81
11/27/2006	12	--	25.39	374.73	--	41.53
11/28/2006	12	--	25.39	349.06	--	38.75
11/29/2006	12	--	25.39	329.13	--	36.60
11/30/2006	12	--	25.39	312.63	--	34.81
12/1/2006	12	--	25.39	359.70	--	39.91

Appendix 1-B.   Measured and MOVE.1-estimated daily mean streamflow at McTier Creek near New Holland, S.C. (station 02172305) and Fishing Brook (County Line Flow) near Newcomb, N.Y. (station 0131199050).—Continued

[--, no data MOVE.1 regression technique used measured daily mean streamflow at long-term index station McTier Creek near Monetta, S.C. (station 02172300). MOVE.1 regression technique used measured daily mean streamflow at long-term index station Hudson River near Newcomb, N.Y. (station 01312000).]

	Daily mean streamflow, in cubic feet per second					
	McTier Creek			Fishing Brook		
Date	Measured for MOVE.1 long-term index station 02172300	Measured for 02172305	MOVE.1 estimate for 02172305	Measured at MOVE.1 long-term index station 01312000	Measured at 0131199050	MOVE.1 estimate for 013199050
12/2/2006	11	--	23.48	841.05	--	91.35
12/3/2006	11	--	23.48	990.85	--	107.18
12/4/2006	11	--	23.48	769.01	--	83.71
12/5/2006	11	--	23.48	607.58	--	66.53
12/6/2006	11	--	23.48	508.51	--	55.93
12/7/2006	11	--	23.48	452.08	--	49.87
12/8/2006	10	--	21.56	389.43	--	43.12
12/9/2006	10	--	21.56	351.73	--	39.04
12/10/2006	10	--	21.56	328.44	--	36.52
12/11/2006	11	--	23.48	313.09	--	34.86
12/12/2006	11	--	23.48	297.21	--	33.13
12/13/2006	11	--	23.48	285.11	--	31.82
12/14/2006	12	--	25.39	293.04	--	32.68
12/15/2006	11	--	23.48	310.18	--	34.54
12/16/2006	11	--	23.48	337.23	--	37.47
12/17/2006	11	--	23.48	359.33	--	39.87
12/18/2006	11	--	23.48	361.94	--	40.15
12/19/2006	11	--	23.48	356.12	--	39.52
12/20/2006	11	--	23.48	332.71	--	36.98
12/21/2006	12	--	25.39	312.66	--	34.81
12/22/2006	90	--	154.18	292.98	--	32.67
12/23/2006	99	--	167.92	359.58	--	39.89
12/24/2006	31	--	59.38	762.78	--	83.05
12/25/2006	57	--	102.43	833.02	--	90.50
12/26/2006	43	--	79.59	745.64	--	81.23
12/27/2006	26	--	50.73	725.41	--	79.08
12/28/2006	22	--	43.68	650.52	--	71.11
12/29/2006	19	--	38.31	532.22	--	58.47
12/30/2006	19	--	38.31	471.31	--	51.94
12/31/2006	17	--	34.68	427.20	--	47.19
1/1/2007	18	--	36.50	413.48	--	45.71
1/2/2007	20	--	40.11	432.55	--	47.77
1/3/2007	17	--	34.68	423.29	--	46.77
1/4/2007	16	--	32.84	397.58	--	44.00
1/5/2007	18	--	36.50	376.80	--	41.76
1/6/2007	34	--	64.50	878.15	--	95.28

Appendix 1-B.   Measured and MOVE.1-estimated daily mean streamflow at McTier Creek near New Holland, S.C. (station 02172305) and Fishing Brook (County Line Flow) near Newcomb, N.Y. (station 0131199050).—Continued

[--, no data MOVE.1 regression technique used measured daily mean streamflow at long-term index station McTier Creek near Monetta, S.C. (station 02172300). MOVE.1 regression technique used measured daily mean streamflow at long-term index station Hudson River near Newcomb, N.Y. (station 01312000).]

| Date | Daily mean streamflow, in cubic feet per second | | | | | |
| | McTier Creek | | | Fishing Brook | | |
	Measured for MOVE.1 long-term index station 02172300	Measured for 02172305	MOVE.1 estimate for 02172305	Measured at MOVE.1 long-term index station 01312000	Measured at 0131199050	MOVE.1 estimate for 013199050
1/7/2007	26	--	50.73	2,833.00	--	298.50
1/8/2007	34	--	64.50	2,767.65	--	291.78
1/9/2007	24	--	47.22	2,367.65	--	250.59
1/10/2007	19	--	38.31	1,823.56	--	194.27
1/11/2007	17	--	34.68	1,300.98	--	139.77
1/12/2007	16	--	32.84	1,074.10	--	115.95
1/13/2007	16	--	32.84	931.88	--	100.95
1/14/2007	16	--	32.84	889.38	--	96.46
1/15/2007	15	--	31.00	829.95	--	90.17
1/16/2007	15	--	31.00	775.91	--	84.44
1/17/2007	15	--	31.00	615.58	--	67.38
1/18/2007	18	--	36.50	523.50	--	57.54
1/19/2007	19	--	38.31	489.14	--	53.85
1/20/2007	16	--	32.84	442.60	--	48.85
1/21/2007	15	--	31.00	384.96	--	42.64
1/22/2007	55	--	99.21	356.76	--	39.59
1/23/2007	32	--	61.09	335.32	--	37.27
1/24/2007	21	--	41.90	314.83	--	35.04
1/25/2007	18	--	36.50	289.33	36.02	32.27
1/26/2007	17	--	34.68	265.44	31.92	29.67
1/27/2007	16	--	32.84	249.09	30.27	27.89
1/28/2007	18	--	36.50	240.19	30.33	26.92
1/29/2007	16	--	32.84	229.79	29.52	25.78
1/30/2007	16	--	32.84	217.89	27.56	24.48
1/31/2007	15	--	31.00	206.64	26.46	23.25
2/1/2007	30	--	57.66	197.74	25.99	22.27
2/2/2007	35	--	66.19	191.96	26.05	21.63
2/3/2007	21	--	41.90	188.09	26.56	21.21
2/4/2007	18	--	36.50	182.90	25.29	20.64
2/5/2007	16	--	32.84	176.04	24.31	19.88
2/6/2007	15	--	31.00	169.78	22.36	19.19
2/7/2007	15	--	31.00	162.87	21.29	18.43
2/8/2007	15	--	31.00	160.15	21.13	18.13
2/9/2007	15	--	31.00	159.23	21.48	18.03
2/10/2007	14	--	29.14	156.97	21.28	17.78
2/11/2007	14	--	29.14	154.71	21.21	17.53

Appendix 1-B.   Measured and MOVE.1-estimated daily mean streamflow at McTier Creek near New Holland, S.C. (station 02172305) and Fishing Brook (County Line Flow) near Newcomb, N.Y. (station 0131199050).—Continued

[--, no data MOVE.1 regression technique used measured daily mean streamflow at long-term index station McTier Creek near Monetta, S.C. (station 02172300). MOVE.1 regression technique used measured daily mean streamflow at long-term index station Hudson River near Newcomb, N.Y. (station 01312000).]

| Date | Daily mean streamflow, in cubic feet per second | | | | | |
| | McTier Creek | | | Fishing Brook | | |
	Measured for MOVE.1 long-term index station 02172300	Measured for 02172305	MOVE.1 estimate for 02172305	Measured at MOVE.1 long-term index station 01312000	Measured at 0131199050	MOVE.1 estimate for 013199050
2/12/2007	14	--	29.14	152.45	21.00	17.28
2/13/2007	16	--	32.84	147.92	21.09	16.78
2/14/2007	30	--	57.66	168.54	25.93	19.06
2/15/2007	18	--	36.50	228.85	26.90	25.68
2/16/2007	16	--	32.84	257.09	26.73	28.76
2/17/2007	15	--	31.00	268.23	25.77	29.98
2/18/2007	15	--	31.00	252.41	25.06	28.25
2/19/2007	14	--	29.14	229.49	22.39	25.75
2/20/2007	14	--	29.14	212.08	22.57	23.84
2/21/2007	25	--	48.98	197.38	22.59	22.23
2/22/2007	29	--	55.94	181.58	22.76	20.49
2/23/2007	20	--	40.11	169.28	21.91	19.14
2/24/2007	16	--	32.84	158.30	21.58	17.93
2/25/2007	17	--	34.68	152.33	22.08	17.27
2/26/2007	17	--	34.68	148.37	22.49	16.83
2/27/2007	14	--	29.14	144.32	21.96	16.38
2/28/2007	14	--	29.14	138.33	20.82	15.72
3/1/2007	33	--	62.80	133.16	20.16	15.14
3/2/2007	173	--	276.78	141.49	23.54	16.07
3/3/2007	68	--	119.96	147.37	24.53	16.72
3/4/2007	36	--	67.88	145.12	23.47	16.47
3/5/2007	28	--	54.21	141.70	21.97	16.09
3/6/2007	24	--	47.22	136.89	20.41	15.56
3/7/2007	22	--	43.68	134.50	20.10	15.29
3/8/2007	20	--	40.11	133.46	19.44	15.18
3/9/2007	19	--	38.31	132.92	18.98	15.12
3/10/2007	18	--	36.50	132.77	19.46	15.10
3/11/2007	17	--	34.68	132.38	21.02	15.06
3/12/2007	17	--	34.68	129.65	21.30	14.75
3/13/2007	17	--	34.68	129.05	21.95	14.69
3/14/2007	17	--	34.68	139.65	29.12	15.86
3/15/2007	17	--	34.68	182.16	67.61	20.56
3/16/2007	20	--	40.11	289.47	104.11	32.29
3/17/2007	19	--	38.31	541.60	100.20	59.47
3/18/2007	16	--	32.84	619.84	83.82	67.84
3/19/2007	15	--	31.00	535.70	72.05	58.84

Appendix 1-B.   Measured and MOVE.1-estimated daily mean streamflow at McTier Creek near New Holland, S.C. (station 02172305) and Fishing Brook (County Line Flow) near Newcomb, N.Y. (station 0131199050).—Continued

[--, no data MOVE.1 regression technique used measured daily mean streamflow at long-term index station McTier Creek near Monetta, S.C. (station 02172300). MOVE.1 regression technique used measured daily mean streamflow at long-term index station Hudson River near Newcomb, N.Y. (station 01312000).]

	Daily mean streamflow, in cubic feet per second					
	McTier Creek			**Fishing Brook**		
Date	**Measured for MOVE.1 long-term index station 02172300**	**Measured for 02172305**	**MOVE.1 estimate for 02172305**	**Measured at MOVE.1 long-term index station 01312000**	**Measured at 0131199050**	**MOVE.1 estimate for 013199050**
3/20/2007	16	--	32.84	446.35	59.24	49.25
3/21/2007	15	--	31.00	376.68	46.59	41.74
3/22/2007	15	--	31.00	331.54	43.91	36.86
3/23/2007	14	--	29.14	338.39	68.57	37.60
3/24/2007	14	--	29.14	427.58	93.51	47.23
3/25/2007	14	--	29.14	544.38	96.87	59.77
3/26/2007	13	--	27.27	591.38	97.49	64.80
3/27/2007	13	--	27.27	711.87	130.77	77.64
3/28/2007	13	--	27.27	1,115.97	218.06	120.35
3/29/2007	13	--	27.27	1,530.81	208.33	163.79
3/30/2007	13	--	27.27	1,432.05	160.20	153.48
3/31/2007	13	--	27.27	1,209.89	132.65	130.22
4/1/2007	13	--	27.27	1,064.20	120.18	114.91
4/2/2007	13	--	27.27	1,012.12	123.26	109.42
4/3/2007	13	--	27.27	1,037.29	144.33	112.07
4/4/2007	12	--	25.39	1,210.09	174.05	130.24
4/5/2007	11	--	23.48	1,316.91	164.89	141.44
4/6/2007	11	--	23.48	1,151.81	124.51	124.12
4/7/2007	11	--	23.48	969.39	98.63	104.91
4/8/2007	11	--	23.48	825.76	81.84	89.73
4/9/2007	11	--	23.48	705.91	68.08	77.01
4/10/2007	11	--	23.48	586.07	58.87	64.23
4/11/2007	13	--	27.27	494.26	52.01	54.40
4/12/2007	26	--	50.73	468.23	58.46	51.61
4/13/2007	16	--	32.84	454.51	58.07	50.13
4/14/2007	13	--	27.27	419.23	51.64	46.33
4/15/2007	26	--	50.73	400.29	51.57	44.29
4/16/2007	19	--	38.31	482.80	69.35	53.17
4/17/2007	14	--	29.14	863.99	116.10	93.78
4/18/2007	12	--	25.39	1,003.54	125.39	108.52
4/19/2007	12	--	25.39	1,116.31	139.75	120.39
4/20/2007	12	--	25.39	1,466.59	189.32	157.09
4/21/2007	12	--	25.39	1,874.26	244.48	199.53
4/22/2007	11	--	23.48	2,356.98	316.06	249.49
4/23/2007	10	--	21.56	2,898.11	406.08	305.18
4/24/2007	10	--	21.56	4,101.01	629.64	428.12

Appendix 1-B. Measured and MOVE.1-estimated daily mean streamflow at McTier Creek near New Holland, S.C. (station 02172305) and Fishing Brook (County Line Flow) near Newcomb, N.Y. (station 0131199050).—Continued

[--, no data MOVE.1 regression technique used measured daily mean streamflow at long-term index station McTier Creek near Monetta, S.C. (station 02172300). MOVE.1 regression technique used measured daily mean streamflow at long-term index station Hudson River near Newcomb, N.Y. (station 01312000).]

| Date | Daily mean streamflow, in cubic feet per second | | | | | |
| | McTier Creek | | | Fishing Brook | | |
	Measured for MOVE.1 long-term index station 02172300	Measured for 02172305	MOVE.1 estimate for 02172305	Measured at MOVE.1 long-term index station 01312000	Measured at 0131199050	MOVE.1 estimate for 013199050
4/25/2007	9.7	--	20.98	4,492.90	449.53	467.97
4/26/2007	9.3	--	20.21	3,063.69	257.38	322.17
4/27/2007	8.9	--	19.43	2,217.24	196.51	235.05
4/28/2007	9.4	--	20.40	1,937.21	200.93	206.06
4/29/2007	10	--	21.56	1,902.15	201.69	202.42
4/30/2007	10	--	21.56	1,878.08	204.75	199.93
5/1/2007	9.1	--	19.82	1,805.28	186.47	192.37
5/2/2007	8.7	--	19.03	1,572.71	150.21	168.16
5/3/2007	8.4	--	18.45	1,356.91	123.99	145.63
5/4/2007	8.3	--	18.25	1,186.91	106.04	127.81
5/5/2007	9.9	--	21.37	1,050.95	91.16	113.51
5/6/2007	11	--	23.48	945.84	78.85	102.43
5/7/2007	9.5	--	20.59	845.60	69.27	91.83
5/8/2007	8.7	--	19.03	799.26	63.59	86.92
5/9/2007	8.6	--	18.84	876.57	63.98	95.11
5/10/2007	8.4	--	18.45	1,030.07	74.43	111.31
5/11/2007	8.1	--	17.86	1,281.08	128.59	137.69
5/12/2007	25	--	48.98	1,336.35	123.10	143.47
5/13/2007	38	--	71.25	1,098.54	88.72	118.52
5/14/2007	14	--	29.14	857.62	65.95	93.10
5/15/2007	10	--	21.56	728.25	57.44	79.38
5/16/2007	9.2	--	20.01	1,015.13	89.48	109.74
5/17/2007	8.7	--	19.03	1,986.51	134.38	211.17
5/18/2007	8.4	--	18.45	1,817.45	112.61	193.63
5/19/2007	7.9	--	17.46	1,277.26	85.41	137.28
5/20/2007	7.3	--	16.27	987.26	71.43	106.80
5/21/2007	7.2	--	16.07	886.94	62.90	96.20
5/22/2007	6.8	--	15.27	791.12	53.88	86.06
5/23/2007	6.6	--	14.86	660.35	45.44	72.16
5/24/2007	6.3	--	14.26	565.78	39.02	62.06
5/25/2007	5.9	--	13.44	526.71	33.17	57.88
5/26/2007	5.8	--	13.24	484.01	29.08	53.30
5/27/2007	5.6	--	12.83	420.69	26.21	46.49
5/28/2007	5.5	--	12.63	393.51	29.89	43.56
5/29/2007	5.3	--	12.21	376.09	27.94	41.68
5/30/2007	4.9	--	11.38	322.04	23.59	35.83

Appendix 1-B. Measured and MOVE.1-estimated daily mean streamflow at McTier Creek near New Holland, S.C. (station 02172305) and Fishing Brook (County Line Flow) near Newcomb, N.Y. (station 0131199050).—Continued

[--, no data MOVE.1 regression technique used measured daily mean streamflow at long-term index station McTier Creek near Monetta, S.C. (station 02172300). MOVE.1 regression technique used measured daily mean streamflow at long-term index station Hudson River near Newcomb, N.Y. (station 01312000).]

| | Daily mean streamflow, in cubic feet per second | | | | | |
| | McTier Creek | | | Fishing Brook | | |
Date	Measured for MOVE.1 long-term index station 02172300	Measured for 02172305	MOVE.1 estimate for 02172305	Measured at MOVE.1 long-term index station 01312000	Measured at 0131199050	MOVE.1 estimate for 013199050
5/31/2007	4.7	--	10.97	278.48	20.73	31.09
6/1/2007	4.8	--	11.18	248.71	18.81	27.85
6/2/2007	5.8	--	13.24	225.18	16.60	25.28
6/3/2007	59	--	105.64	202.71	14.81	22.81
6/4/2007	24	--	47.22	190.80	22.97	21.51
6/5/2007	12	--	25.39	209.20	30.92	23.53
6/6/2007	8.5	--	18.64	224.87	25.92	25.24
6/7/2007	7.1	--	15.87	203.80	20.54	22.93
6/8/2007	5.9	--	13.44	181.59	17.40	20.49
6/9/2007	5.2	--	12.01	170.17	15.39	19.24
6/10/2007	4.6	--	10.76	152.12	12.40	17.24
6/11/2007	4.9	--	11.38	136.57	11.27	15.52
6/12/2007	7.5	--	16.67	123.48	10.00	14.07
6/13/2007	6.5	19	14.66	110.12	8.97	12.58
6/14/2007	7.8	21	17.26	99.06	8.01	11.35
6/15/2007	8.4	27	18.45	89.98	7.26	10.33
6/16/2007	8.8	29	19.23	82.80	6.84	9.53
6/17/2007	6.9	20	15.47	77.80	7.11	8.97
6/18/2007	5.9	17	13.44	73.90	7.70	8.53
6/19/2007	4.8	14	11.18	72.48	7.21	8.37
6/20/2007	4.4	13	10.34	109.71	27.75	12.54
6/21/2007	4.8	14	11.18	147.49	43.11	16.73
6/22/2007	4.4	13	10.34	153.46	40.13	17.39
6/23/2007	3.9	12	9.28	157.32	34.42	17.82
6/24/2007	3.5	11	8.42	158.39	26.54	17.94
6/25/2007	3.6	11	8.64	144.22	20.04	16.37
6/26/2007	6.4	13	14.46	128.90	15.03	14.67
6/27/2007	4.5	11	10.55	115.78	11.65	13.21
6/28/2007	4	9.9	9.49	105.58	9.54	12.08
6/29/2007	3.7	9.7	8.85	93.91	7.15	10.77
6/30/2007	3.4	9.3	8.21	83.86	6.47	9.65
7/1/2007	3.5	9.4	8.42	77.22	6.41	8.90
7/2/2007	28	49	54.21	71.65	8.07	8.28
7/3/2007	11	53	23.48	66.18	5.94	7.66
7/4/2007	6.5	20	14.66	63.25	6.29	7.33
7/5/2007	5.6	15	12.83	71.32	8.26	8.24

Appendix 1-B.   Measured and MOVE.1-estimated daily mean streamflow at McTier Creek near New Holland, S.C. (station 02172305) and Fishing Brook (County Line Flow) near Newcomb, N.Y. (station 0131199050).—Continued

[--, no data MOVE.1 regression technique used measured daily mean streamflow at long-term index station McTier Creek near Monetta, S.C. (station 02172300). MOVE.1 regression technique used measured daily mean streamflow at long-term index station Hudson River near Newcomb, N.Y. (station 01312000).]

| Date | Daily mean streamflow, in cubic feet per second | | | | | |
| | McTier Creek | | | Fishing Brook | | |
	Measured for MOVE.1 long-term index station 02172300	Measured for 02172305	MOVE.1 estimate for 02172305	Measured at MOVE.1 long-term index station 01312000	Measured at 0131199050	MOVE.1 estimate for 013199050
7/6/2007	5	13	11.59	90.96	9.24	10.44
7/7/2007	4.7	12	10.97	113.16	9.89	12.92
7/8/2007	5.1	13	11.80	142.19	8.61	16.15
7/9/2007	5.5	15	12.63	202.67	12.38	22.81
7/10/2007	5.7	14	13.04	269.25	15.37	30.09
7/11/2007	7.8	23	17.26	247.79	16.14	27.75
7/12/2007	7.7	30	17.06	274.37	25.62	30.65
7/13/2007	6.2	19	14.05	314.02	25.47	34.96
7/14/2007	5.6	15	12.83	261.75	17.55	29.27
7/15/2007	5.6	15	12.83	217.26	14.50	24.41
7/16/2007	5.2	16	12.01	186.90	12.22	21.08
7/17/2007	4.7	13	10.97	160.19	10.08	18.14
7/18/2007	4.4	12	10.34	142.18	9.15	16.14
7/19/2007	4.2	12	9.92	142.63	12.57	16.19
7/20/2007	3.9	11	9.28	273.79	30.78	30.58
7/21/2007	3.8	10	9.07	402.90	55.98	44.57
7/22/2007	3.6	9.5	8.64	374.10	52.40	41.46
7/23/2007	3.2	8.9	7.77	296.15	30.30	33.02
7/24/2007	3	8.3	7.34	238.61	19.39	26.74
7/25/2007	3.1	8.2	7.56	197.63	13.78	22.26
7/26/2007	2.9	8.2	7.12	166.03	10.79	18.78
7/27/2007	2.9	8.5	7.12	145.17	8.52	16.48
7/28/2007	3.5	9.6	8.42	135.09	8.05	15.36
7/29/2007	3.8	12	9.07	126.65	8.54	14.42
7/30/2007	3.7	13	8.85	112.12	7.97	12.81
7/31/2007	3.8	15	9.07	98.35	6.83	11.27
8/1/2007	3.4	12	8.21	86.73	6.10	9.97
8/2/2007	3.3	10	7.99	77.39	5.41	8.92
8/3/2007	2.9	9	7.12	69.31	4.86	8.01
8/4/2007	2.8	8.1	6.90	63.26	5.35	7.33
8/5/2007	2.7	7.3	6.68	56.43	5.38	6.56
8/6/2007	2.6	6.6	6.46	54.01	5.42	6.28
8/7/2007	2.5	6.3	6.23	56.75	5.77	6.59
8/8/2007	2.4	6.2	6.01	67.89	9.12	7.85
8/9/2007	2.3	5.6	5.78	83.63	9.73	9.62
8/10/2007	2.2	5.3	5.56	83.13	8.83	9.57

Appendix 1-B.   Measured and MOVE.1-estimated daily mean streamflow at McTier Creek near New Holland, S.C. (station 02172305) and Fishing Brook (County Line Flow) near Newcomb, N.Y. (station 0131199050).—Continued

[--, no data MOVE.1 regression technique used measured daily mean streamflow at long-term index station McTier Creek near Monetta, S.C. (station 02172300). MOVE.1 regression technique used measured daily mean streamflow at long-term index station Hudson River near Newcomb, N.Y. (station 01312000).]

| | Daily mean streamflow, in cubic feet per second | | | | | |
| | McTier Creek | | | Fishing Brook | | |
Date	Measured for MOVE.1 long-term index station 02172300	Measured for 02172305	MOVE.1 estimate for 02172305	Measured at MOVE.1 long-term index station 01312000	Measured at 0131199050	MOVE.1 estimate for 013199050
8/11/2007	2.1	4.7	5.33	75.05	7.70	8.66
8/12/2007	2.1	4.5	5.33	67.54	5.86	7.81
8/13/2007	2.2	5	5.56	60.44	4.86	7.01
8/14/2007	2	4.6	5.10	53.95	4.85	6.28
8/15/2007	2	4.3	5.10	48.64	4.15	5.67
8/16/2007	2.1	4.5	5.33	45.35	3.13	5.30
8/17/2007	2.1	5.5	5.33	45.02	3.37	5.26
8/18/2007	2	5.1	5.10	45.33	4.21	5.30
8/19/2007	2	4.7	5.10	45.27	4.03	5.29
8/20/2007	2	4.4	5.10	42.36	4.38	4.96
8/21/2007	2	4.2	5.10	39.27	4.12	4.61
8/22/2007	1.9	4.1	4.87	36.29	3.68	4.26
8/23/2007	2.2	5	5.56	34.66	3.84	4.08
8/24/2007	2.2	5.1	5.56	35.40	4.62	4.16
8/25/2007	2.1	5	5.33	35.77	4.58	4.20
8/26/2007	3.5	12	8.42	34.20	4.26	4.02
8/27/2007	5.4	18	12.42	32.27	3.95	3.80
8/28/2007	4	12	9.49	28.87	3.53	3.41
8/29/2007	4.4	10	10.34	26.45	3.32	3.13
8/30/2007	8	27	17.66	25.05	3.14	2.97
8/31/2007	4.9	17	11.38	24.06	2.86	2.86
9/1/2007	4.3	14	10.13	22.68	2.50	2.70
9/2/2007	4	12	9.49	21.69	2.25	2.58
9/3/2007	3.4	9.7	8.21	21.00	2.28	2.50
9/4/2007	3.2	8.8	7.77	20.14	2.23	2.40
9/5/2007	3	8.2	7.34	19.37	2.08	2.31
9/6/2007	2.9	7.6	7.12	18.70	2.26	2.23
9/7/2007	2.8	7.1	6.90	18.00	2.54	2.15
9/8/2007	2.7	6.9	6.68	17.76	3.46	2.12
9/9/2007	2.6	6.5	6.46	28.41	4.66	3.36
9/10/2007	2.4	6	6.01	41.66	6.47	4.88
9/11/2007	2.5	5.8	6.23	38.70	7.75	4.54
9/12/2007	2.6	6.9	6.46	39.62	10.45	4.64
9/13/2007	3.3	11	7.99	39.57	10.01	4.64
9/14/2007	3.9	11	9.28	36.81	8.63	4.32
9/15/2007	8	24	17.66	38.24	11.21	4.49

Appendix 1-B. Measured and MOVE.1-estimated daily mean streamflow at McTier Creek near New Holland, S.C. (station 02172305) and Fishing Brook (County Line Flow) near Newcomb, N.Y. (station 0131199050).—Continued

[--, no data MOVE.1 regression technique used measured daily mean streamflow at long-term index station McTier Creek near Monetta, S.C. (station 02172300). MOVE.1 regression technique used measured daily mean streamflow at long-term index station Hudson River near Newcomb, N.Y. (station 01312000).]

Date	Daily mean streamflow, in cubic feet per second					
	McTier Creek			Fishing Brook		
	Measured for MOVE.1 long-term index station 02172300	Measured for 02172305	MOVE.1 estimate for 02172305	Measured at MOVE.1 long-term index station 01312000	Measured at 0131199050	MOVE.1 estimate for 013199050
9/16/2007	5.5	16	12.63	40.57	10.76	4.75
9/17/2007	4.1	12	9.71	41.50	10.05	4.86
9/18/2007	3.6	10	8.64	39.96	8.80	4.68
9/19/2007	3.4	9.8	8.21	38.20	7.26	4.48
9/20/2007	3.4	9.7	8.21	36.46	5.85	4.28
9/21/2007	4	11	9.49	34.88	4.64	4.10
9/22/2007	3.9	12	9.28	33.06	3.84	3.89
9/23/2007	3.6	11	8.64	31.50	3.37	3.71
9/24/2007	3.4	9.8	8.21	29.26	2.58	3.46
9/25/2007	3.2	9	7.77	27.43	2.67	3.24
9/26/2007	3.1	8.5	7.56	28.85	4.46	3.41
9/27/2007	3.1	8.6	7.56	30.90	4.77	3.65
9/28/2007	3	8.7	7.34	48.62	11.23	5.67
9/29/2007	2.8	8	6.90	65.73	14.14	7.61
9/30/2007	2.8	6.9	6.90	67.65	12.66	7.83
10/1/2007	2.8	7.1	6.90	60.68	10.04	7.04
10/2/2007	2.8	6.9	6.90	53.90	7.61	6.27
10/3/2007	3.3	8.4	7.99	48.82	6.19	5.69
10/4/2007	4.2	11	9.92	45.09	5.48	5.27
10/5/2007	5.3	15	12.21	41.79	5.04	4.89
10/6/2007	4.6	13	10.76	40.40	4.74	4.73
10/7/2007	4.5	12	10.55	41.83	4.98	4.90
10/8/2007	4.3	12	10.13	61.45	10.90	7.13
10/9/2007	4	11	9.49	79.05	23.60	9.11
10/10/2007	3.8	9.7	9.07	115.51	28.66	13.18
10/11/2007	3.7	9.1	8.85	172.28	28.27	19.47
10/12/2007	3.5	8.6	8.42	176.69	30.72	19.95
10/13/2007	3.4	8.6	8.21	178.43	31.93	20.15
10/14/2007	3.4	8.5	8.21	180.64	33.74	20.39
10/15/2007	3.4	8.4	8.21	176.95	31.07	19.98
10/16/2007	3.3	8.4	7.99	161.33	24.78	18.26
10/17/2007	3.5	9	8.42	152.66	19.09	17.30
10/18/2007	3.4	9.1	8.21	142.46	15.95	16.18
10/19/2007	3.8	10	9.07	138.24	17.15	15.71
10/20/2007	3.9	11	9.28	511.73	63.21	56.27
10/21/2007	3.4	9.2	8.21	896.31	84.49	97.20

Appendix 1-B. Measured and MOVE.1-estimated daily mean streamflow at McTier Creek near New Holland, S.C. (station 02172305) and Fishing Brook (County Line Flow) near Newcomb, N.Y. (station 0131199050).—Continued

[--, no data MOVE.1 regression technique used measured daily mean streamflow at long-term index station McTier Creek near Monetta, S.C. (station 02172300). MOVE.1 regression technique used measured daily mean streamflow at long-term index station Hudson River near Newcomb, N.Y. (station 01312000).]

	Daily mean streamflow, in cubic feet per second					
	McTier Creek			Fishing Brook		
Date	Measured for MOVE.1 long-term index station 02172300	Measured for 02172305	MOVE.1 estimate for 02172305	Measured at MOVE.1 long-term index station 01312000	Measured at 0131199050	MOVE.1 estimate for 013199050
10/22/2007	3.4	8.8	8.21	605.19	54.43	66.27
10/23/2007	3.8	9.4	9.07	475.67	53.59	52.40
10/24/2007	6.8	18	15.27	1,250.20	159.75	134.45
10/25/2007	11	35	23.48	1,495.88	126.12	160.15
10/26/2007	7.6	20	16.87	1,040.60	74.05	112.42
10/27/2007	6.4	16	14.46	839.67	66.94	91.20
10/28/2007	5.7	14	13.04	1,307.98	114.18	140.50
10/29/2007	5.4	12	12.42	1,350.32	105.31	144.94
10/30/2007	5.4	12	12.42	984.64	73.59	106.52
10/31/2007	5.5	12	12.63	731.80	52.56	79.76
11/1/2007	5.5	12	12.63	543.59	41.63	59.69
11/2/2007	5.5	12	12.63	424.67	33.59	46.92
11/3/2007	5.4	11	12.42	343.61	29.08	38.16
11/4/2007	5.5	11	12.63	290.18	25.26	32.37
11/5/2007	5.4	11	12.42	249.15	22.06	27.90
11/6/2007	5.5	11	12.63	244.92	28.18	27.43
11/7/2007	5	10	11.59	283.52	36.62	31.64
11/8/2007	5.3	10	12.21	276.66	36.17	30.90
11/9/2007	6.1	12	13.85	248.58	31.23	27.83
11/10/2007	6.4	12	14.46	223.80	26.73	25.13
11/11/2007	6.4	13	14.46	201.66	23.02	22.70
11/12/2007	6.7	13	15.07	184.06	20.89	20.76
11/13/2007	7	14	15.67	176.42	22.20	19.92
11/14/2007	7	14	15.67	175.38	23.24	19.81
11/15/2007	7.8	15	17.26	239.51	39.34	26.84
11/16/2007	8.4	15	18.45	609.33	75.29	66.72
11/17/2007	7.9	15	17.46	676.76	73.94	73.91
11/18/2007	7.7	14	17.06	529.19	55.48	58.15
11/19/2007	7.8	15	17.26	413.29	46.37	45.69
11/20/2007	8.2	15	18.05	362.51	41.61	40.21
11/21/2007	8.6	15	18.84	334.63	40.88	37.19
11/22/2007	9	15	19.62	513.84	97.82	56.50
11/23/2007	10	18	21.56	1,175.53	176.05	126.61
11/24/2007	9.6	17	20.79	1,157.82	124.42	124.75
11/25/2007	8.9	16	19.43	893.69	89.08	96.92
11/26/2007	9.3	16	20.21	724.94	67.82	79.03

Appendix 1-B.   Measured and MOVE.1-estimated daily mean streamflow at McTier Creek near New Holland, S.C. (station 02172305) and Fishing Brook (County Line Flow) near Newcomb, N.Y. (station 0131199050).—Continued

[--, no data MOVE.1 regression technique used measured daily mean streamflow at long-term index station McTier Creek near Monetta, S.C. (station 02172300). MOVE.1 regression technique used measured daily mean streamflow at long-term index station Hudson River near Newcomb, N.Y. (station 01312000).]

Date	Daily mean streamflow, in cubic feet per second					
	McTier Creek			Fishing Brook		
	Measured for MOVE.1 long-term index station 02172300	Measured for 02172305	MOVE.1 estimate for 02172305	Measured at MOVE.1 long-term index station 01312000	Measured at 0131199050	MOVE.1 estimate for 013199050
11/27/2007	10	18	21.56	769.99	104.64	83.81
11/28/2007	9.9	17	21.37	980.03	133.80	106.04
11/29/2007	9.6	16	20.79	892.87	103.43	96.83
11/30/2007	9.4	16	20.40	771.90	78.74	84.02
12/1/2007	9.3	16	20.21	613.62	60.19	67.17
12/2/2007	9.3	15	20.21	485.04	47.74	53.41
12/3/2007	9.7	15	20.98	432.73	46.39	47.79
12/4/2007	9.4	15	20.40	394.04	42.95	43.62
12/5/2007	8.7	14	19.03	352.61	39.43	39.14
12/6/2007	10	14	21.56	307.99	34.81	34.30
12/7/2007	10	16	21.56	279.42	31.70	31.20
12/8/2007	10	17	21.56	258.28	29.94	28.89
12/9/2007	11	16	23.48	235.40	28.12	26.39
12/10/2007	10	17	21.56	220.81	27.21	24.80
12/11/2007	9.9	17	21.37	209.83	26.84	23.59
12/12/2007	10	16	21.56	214.60	31.25	24.12
12/13/2007	11	15	23.48	208.37	32.16	23.44
12/14/2007	11	16	23.48	206.40	32.18	23.22
12/15/2007	13	19	27.27	198.91	28.35	22.40
12/16/2007	60	90	107.25	201.94	30.88	22.73
12/17/2007	19	51	38.31	203.90	31.15	22.94
12/18/2007	13	26	27.27	201.91	29.90	22.73
12/19/2007	12	21	25.39	200.24	27.72	22.54
12/20/2007	12	19	25.39	192.72	26.14	21.72
12/21/2007	18	34	36.50	183.56	24.83	20.71
12/22/2007	16	32	32.84	173.77	24.41	19.63
12/23/2007	14	24	29.14	171.50	28.16	19.38
12/24/2007	13	23	27.27	291.46	68.29	32.51
12/25/2007	13	20	27.27	654.83	103.95	71.57
12/26/2007	17	33	34.68	708.91	95.98	77.33
12/27/2007	15	28	31.00	592.51	75.42	64.92
12/28/2007	14	23	29.14	500.84	59.38	55.11
12/29/2007	15	28	31.00	448.76	61.53	49.51
12/30/2007	49	76	89.46	404.87	70.00	44.78
12/31/2007	42	91	77.93	377.90	64.91	41.87
1/1/2008	20	43	40.11	351.92	55.09	39.06

Appendix 1-B. Measured and MOVE.1-estimated daily mean streamflow at McTier Creek near New Holland, S.C. (station 02172305) and Fishing Brook (County Line Flow) near Newcomb, N.Y. (station 0131199050).—Continued

[--, no data MOVE.1 regression technique used measured daily mean streamflow at long-term index station McTier Creek near Monetta, S.C. (station 02172300). MOVE.1 regression technique used measured daily mean streamflow at long-term index station Hudson River near Newcomb, N.Y. (station 01312000).]

| | Daily mean streamflow, in cubic feet per second | | | | | |
| | McTier Creek | | | Fishing Brook | | |
Date	Measured for MOVE.1 long-term index station 02172300	Measured for 02172305	MOVE.1 estimate for 02172305	Measured at MOVE.1 long-term index station 01312000	Measured at 0131199050	MOVE.1 estimate for 013199050
1/2/2008	15	30	31.00	327.37	48.07	36.41
1/3/2008	14	26	29.14	282.48	39.15	31.53
1/4/2008	14	24	29.14	261.25	36.22	29.22
1/5/2008	14	24	29.14	252.96	34.82	28.31
1/6/2008	14	24	29.14	243.46	33.42	27.28
1/7/2008	14	23	29.14	235.65	35.38	26.42
1/8/2008	14	22	29.14	302.10	82.02	33.66
1/9/2008	14	22	29.14	954.95	268.23	103.39
1/10/2008	14	23	29.14	1,797.91	329.96	191.60
1/11/2008	23	47	45.45	1,892.29	202.91	201.40
1/12/2008	21	45	41.90	1,732.06	196.65	184.76
1/13/2008	16	30	32.84	1,490.31	156.15	159.57
1/14/2008	15	25	31.00	1,185.78	114.61	127.69
1/15/2008	14	23	29.14	962.11	92.20	104.15
1/16/2008	14	22	29.14	760.73	68.74	82.83
1/17/2008	30	48	57.66	607.32	57.20	66.50
1/18/2008	26	49	50.73	530.12	53.80	58.25
1/19/2008	29	47	55.94	459.80	49.22	50.70
1/20/2008	32	60	61.09	395.91	43.33	43.82
1/21/2008	20	36	40.11	348.85	37.62	38.73
1/22/2008	18	31	36.50	311.30	36.08	34.66
1/23/2008	20	37	40.11	293.01	35.49	32.67
1/24/2008	18	31	36.50	271.64	33.91	30.35
1/25/2008	16	27	32.84	249.50	31.14	27.93
1/26/2008	16	25	32.84	237.96	30.25	26.67
1/27/2008	16	24	32.84	230.33	28.99	25.84
1/28/2008	15	23	31.00	215.03	28.13	24.16
1/29/2008	15	22	31.00	202.02	26.83	22.74
1/30/2008	18	33	36.50	203.12	31.03	22.86
1/31/2008	16	28	32.84	211.77	35.75	23.81
2/1/2008	27	42	52.47	227.70	35.87	25.55
2/2/2008	23	42	45.45	245.90	38.18	27.54
2/3/2008	18	30	36.50	248.93	37.08	27.87
2/4/2008	16	27	32.84	241.12	32.96	27.02
2/5/2008	16	24	32.84	243.85	39.02	27.32
2/6/2008	18	28	36.50	288.56	74.20	32.19

Appendix 1-B.   Measured and MOVE.1-estimated daily mean streamflow at McTier Creek near New Holland, S.C. (station 02172305) and Fishing Brook (County Line Flow) near Newcomb, N.Y. (station 0131199050).—Continued

[--, no data MOVE.1 regression technique used measured daily mean streamflow at long-term index station McTier Creek near Monetta, S.C. (station 02172300). MOVE.1 regression technique used measured daily mean streamflow at long-term index station Hudson River near Newcomb, N.Y. (station 01312000).]

	Daily mean streamflow, in cubic feet per second					
	McTier Creek			Fishing Brook		
Date	Measured for MOVE.1 long-term index station 02172300	Measured for 02172305	MOVE.1 estimate for 02172305	Measured at MOVE.1 long-term index station 01312000	Measured at 0131199050	MOVE.1 estimate for 013199050
2/7/2008	22	40	43.68	385.86	90.80	42.73
2/8/2008	17	28	34.68	437.95	76.32	48.35
2/9/2008	16	24	32.84	410.17	60.31	45.36
2/10/2008	15	22	31.00	373.09	51.05	41.35
2/11/2008	15	21	31.00	334.22	41.05	37.15
2/12/2008	15	21	31.00	302.18	41.01	33.67
2/13/2008	15	23	31.00	288.78	40.39	32.21
2/14/2008	15	22	31.00	274.08	36.37	30.61
2/15/2008	14	21	29.14	259.08	34.00	28.98
2/16/2008	14	21	29.14	242.42	31.55	27.16
2/17/2008	14	21	29.14	227.03	30.29	25.48
2/18/2008	51	72	92.72	242.17	47.59	27.13
2/19/2008	25	56	48.98	323.30	88.94	35.96
2/20/2008	18	34	36.50	467.46	88.89	51.52
2/21/2008	17	29	34.68	466.18	68.09	51.39
2/22/2008	44	72	81.24	410.56	53.31	45.40
2/23/2008	26	53	50.73	363.41	46.37	40.31
2/24/2008	20	36	40.11	324.40	40.97	36.08
2/25/2008	18	31	36.50	293.55	37.20	32.73
2/26/2008	20	35	40.11	257.06	36.63	28.76
2/27/2008	20	39	40.11	232.24	35.61	26.05
2/28/2008	17	30	34.68	210.00	32.13	23.61
2/29/2008	16	27	32.84	190.00	31.26	21.42
3/1/2008	16	26	32.84	180.00	31.18	20.32
3/2/2008	16	25	32.84	170.00	29.07	19.22
3/3/2008	16	25	32.84	165.00	28.57	18.67
3/4/2008	21	31	41.90	195.00	36.79	21.97
3/5/2008	38	68	71.25	241.36	62.11	27.03
3/6/2008	21	40	41.90	343.74	85.67	38.18
3/7/2008	43	52	79.59	447.83	79.56	49.41
3/8/2008	36	74	67.88	509.30	78.93	56.01
3/9/2008	19	42	38.31	836.10	148.55	90.82
3/10/2008	16	33	32.84	1,263.72	175.29	135.87
3/11/2008	15	30	31.00	1,335.77	141.09	143.41
3/12/2008	14	28	29.14	1,138.81	105.61	122.75
3/13/2008	14	26	29.14	921.00	78.20	99.80

Appendix 1-B. Measured and MOVE.1-estimated daily mean streamflow at McTier Creek near New Holland, S.C. (station 02172305) and Fishing Brook (County Line Flow) near Newcomb, N.Y. (station 0131199050).—Continued

[--, no data MOVE.1 regression technique used measured daily mean streamflow at long-term index station McTier Creek near Monetta, S.C. (station 02172300). MOVE.1 regression technique used measured daily mean streamflow at long-term index station Hudson River near Newcomb, N.Y. (station 01312000).]

Date	Daily mean streamflow, in cubic feet per second					
	McTier Creek			Fishing Brook		
	Measured for MOVE.1 long-term index station 02172300	Measured for 02172305	MOVE.1 estimate for 02172305	Measured at MOVE.1 long-term index station 01312000	Measured at 0131199050	MOVE.1 estimate for 013199050
3/14/2008	13	26	27.27	768.38	64.40	83.64
3/15/2008	13	25	27.27	645.47	59.49	70.57
3/16/2008	14	28	29.14	568.47	59.51	62.35
3/17/2008	13	25	27.27	504.09	54.08	55.46
3/18/2008	13	24	27.27	446.26	48.38	49.24
3/19/2008	14	26	29.14	420.81	49.09	46.50
3/20/2008	22	44	43.68	510.12	73.79	56.10
3/21/2008	15	31	31.00	658.28	83.22	71.94
3/22/2008	13	26	27.27	651.16	68.64	71.18
3/23/2008	13	24	27.27	560.39	55.70	61.49
3/24/2008	12	23	25.39	481.94	47.59	53.08
3/25/2008	12	22	25.39	420.89	42.33	46.51
3/26/2008	12	22	25.39	384.75	41.95	42.61
3/27/2008	12	23	25.39	353.95	39.74	39.28
3/28/2008	12	22	25.39	338.87	40.03	37.65
3/29/2008	12	22	25.39	317.07	36.15	35.29
3/30/2008	12	24	25.39	295.49	33.96	32.94
3/31/2008	12	25	25.39	287.94	35.91	32.12
4/1/2008	12	24	25.39	305.49	55.87	34.03
4/2/2008	12	23	25.39	632.87		69.23
4/3/2008	12	24	25.39	1,119.75	166.62	120.75
4/4/2008	13	26	27.27	1,122.51	136.68	121.04
4/5/2008	32	61	61.09	1,124.15	135.95	121.21
4/6/2008	35	78	66.19	1,235.85	152.55	132.94
4/7/2008	19	43	38.31	1,385.54	184.24	148.62
4/8/2008	15	32	31.00	1,613.72	216.75	172.44
4/9/2008	14	28	29.14	1,900.31	261.21	202.23
4/10/2008	13	26	27.27	2,519.51	380.53	266.25
4/11/2008	12	24	25.39	3,025.44	372.86	318.25
4/12/2008	12	23	25.39	3,123.25	428.63	328.28
4/13/2008	12	24	25.39	3,948.02	508.44	412.54
4/14/2008	12	22	25.39	3,293.27	271.43	345.69
4/15/2008	12	22	25.39	2,303.94	178.02	244.01
4/16/2008	11	21	23.48	1,850.41	164.39	197.05
4/17/2008	11	20	23.48	1,844.15	191.25	196.40
4/18/2008	12	20	25.39	2,301.47	270.22	243.76

Appendix 1-B.   Measured and MOVE.1-estimated daily mean streamflow at McTier Creek near New Holland, S.C. (station 02172305) and Fishing Brook (County Line Flow) near Newcomb, N.Y. (station 0131199050).—Continued

[--, no data MOVE.1 regression technique used measured daily mean streamflow at long-term index station McTier Creek near Monetta, S.C. (station 02172300). MOVE.1 regression technique used measured daily mean streamflow at long-term index station Hudson River near Newcomb, N.Y. (station 01312000).]

| | Daily mean streamflow, in cubic feet per second | | | | | |
| | McTier Creek | | | Fishing Brook | | |
Date	Measured for MOVE.1 long-term index station 02172300	Measured for 02172305	MOVE.1 estimate for 02172305	Measured at MOVE.1 long-term index station 01312000	Measured at 0131199050	MOVE.1 estimate for 013199050
4/19/2008	12	20	25.39	3,008.86	349.58	316.55
4/20/2008	12	21	25.39	3,735.46	380.02	390.87
4/21/2008	12	20	25.39	4,013.57	337.56	419.22
4/22/2008	11	19	23.48	3,682.39	270.76	385.46
4/23/2008	11	19	23.48	3,271.74	222.40	343.49
4/24/2008	11	19	23.48	2,913.01	186.12	306.71
4/25/2008	10	18	21.56	2,524.96	150.16	266.81
4/26/2008	11	20	23.48	2,066.68	120.88	219.48
4/27/2008	12	25	25.39	1,826.30	115.25	194.55
4/28/2008	13	27	27.27	1,748.95	113.55	186.51
4/29/2008	12	26	25.39	1,880.82	144.15	200.21
4/30/2008	10	20	21.56	1,914.62	131.42	203.72
5/1/2008	9.6	18	20.79	1,506.62	100.62	161.27
5/2/2008	9.2	17	20.01	1,137.04	79.38	122.57
5/3/2008	8.8	16	19.23	972.98	72.58	105.29
5/4/2008	8.6	15	18.84	964.16	72.40	104.36
5/5/2008	8.3	14	18.25	1,024.55	72.85	110.73
5/6/2008	8.2	14	18.05	947.14	64.19	102.57
5/7/2008	7.9	14	17.46	871.49	54.36	94.57
5/8/2008	7.7	13	17.06	839.75	56.48	91.21
5/9/2008	7.9	19	17.46	896.96	55.42	97.26
5/10/2008	7.6	15	16.87	798.97	54.80	86.89
5/11/2008	83	71	143.40	666.61	70.17	72.82
5/12/2008	30	94	57.66	589.05	51.94	64.55
5/13/2008	11	32	23.48	527.76	42.98	57.99
5/14/2008	9.3	22	20.21	462.74	36.55	51.02
5/15/2008	8.6	19	18.84	439.05	33.16	48.47
5/16/2008	8.4	18	18.45	433.54	29.59	47.87
5/17/2008	7.8	17	17.26	397.50	27.15	43.99
5/18/2008	7.2	15	16.07	341.75	25.28	37.96
5/19/2008	6.7	14	15.07	330.09	24.68	36.70
5/20/2008	7	13	15.67	329.04	24.59	36.59
5/21/2008	7.3	16	16.27	307.43	23.55	34.24
5/22/2008	6.4	14	14.46	295.14	23.95	32.91
5/23/2008	6.2	13	14.05	289.35	32.20	32.28
5/24/2008	5.9	11	13.44	304.57	39.74	33.93

Appendix 1-B.   Measured and MOVE.1-estimated daily mean streamflow at McTier Creek near New Holland, S.C. (station 02172305) and Fishing Brook (County Line Flow) near Newcomb, N.Y. (station 0131199050).—Continued

[--, no data MOVE.1 regression technique used measured daily mean streamflow at long-term index station McTier Creek near Monetta, S.C. (station 02172300). MOVE.1 regression technique used measured daily mean streamflow at long-term index station Hudson River near Newcomb, N.Y. (station 01312000).]

	Daily mean streamflow, in cubic feet per second					
	McTier Creek			Fishing Brook		
Date	Measured for MOVE.1 long-term index station 02172300	Measured for 02172305	MOVE.1 estimate for 02172305	Measured at MOVE.1 long-term index station 01312000	Measured at 0131199050	MOVE.1 estimate for 013199050
5/25/2008	5.5	11	12.63	313.37	34.11	34.89
5/26/2008	5.1	10	11.80	290.92	28.55	32.45
5/27/2008	4.6	9.6	10.76	311.18	30.90	34.65
5/28/2008	4.8	13	11.18	335.35	32.91	37.27
5/29/2008	5.7	20	13.04	287.01	28.83	32.02
5/30/2008	5	14	11.59	243.72	22.36	27.30
5/31/2008	4.3	11	10.13	229.19	22.73	25.72
6/1/2008	3.9	9.4	9.28	300.43	32.45	33.48
6/2/2008	3.7	8.6	8.85	364.04	35.74	40.38
6/3/2008	3.6	8.6	8.64	353.45	31.08	39.23
6/4/2008	3.3	8	7.99	314.53	27.68	35.01
6/5/2008	3	7.4	7.34	284.50	25.18	31.75
6/6/2008	3	7	7.34	449.98	48.94	49.64
6/7/2008	2.8	6.3	6.90	919.41	74.93	99.64
6/8/2008	2.5	5.8	6.23	834.56	54.33	90.66
6/9/2008	2.4	5.6	6.01	619.92	36.10	67.85
6/10/2008	3.6	6.8	8.64	464.17	26.97	51.17
6/11/2008	7.6	27	16.87	393.62	28.43	43.57
6/12/2008	5.1	15	11.80	332.58	26.06	36.97
6/13/2008	4.2	11	9.92	270.82	19.82	30.26
6/14/2008	3.2	9.3	7.77	227.19	15.83	25.50
6/15/2008	3	8.1	7.34	197.51	13.97	22.24
6/16/2008	3	8.1	7.34	177.48	12.25	20.04
6/17/2008	2.7	7.2	6.68	157.93	10.95	17.89
6/18/2008	2.3	5.8	5.78	142.61	9.81	16.19
6/19/2008	2.2	5.1	5.56	139.88	10.14	15.89
6/20/2008	2.2	4.8	5.56	139.91	10.58	15.89
6/21/2008	2.7	5.1	6.68	146.06	11.65	16.57
6/22/2008	2.3	5.6	5.78	155.96	14.60	17.67
6/23/2008	2.4	5.6	6.01	158.61	17.20	17.96
6/24/2008	2.3	5.4	5.78	175.74	16.41	19.85
6/25/2008	2.1	4.8	5.33	178.01	13.84	20.10
6/26/2008	1.8	4.2	4.64	165.27	12.98	18.70
6/27/2008	2	4.6	5.10	159.21	13.84	18.03
6/28/2008	1.9	4.8	4.87	162.01	16.91	18.34
6/29/2008	1.8	4.1	4.64	264.74	39.53	29.60

Appendix 1-B.   Measured and MOVE.1-estimated daily mean streamflow at McTier Creek near New Holland, S.C. (station 02172305) and Fishing Brook (County Line Flow) near Newcomb, N.Y. (station 0131199050).—Continued

[--, no data MOVE.1 regression technique used measured daily mean streamflow at long-term index station McTier Creek near Monetta, S.C. (station 02172300). MOVE.1 regression technique used measured daily mean streamflow at long-term index station Hudson River near Newcomb, N.Y. (station 01312000).]

Date	Daily mean streamflow, in cubic feet per second					
	McTier Creek			Fishing Brook		
	Measured for MOVE.1 long-term index station 02172300	Measured for 02172305	MOVE.1 estimate for 02172305	Measured at MOVE.1 long-term index station 01312000	Measured at 0131199050	MOVE.1 estimate for 013199050
6/30/2008	2	4.1	5.10	425.55	41.96	47.01
7/1/2008	1.9	4.1	4.87	452.54	28.61	49.92
7/2/2008	1.9	3.4	4.87	344.73	19.91	38.29
7/3/2008	1.9	2.6	4.87	272.54	15.97	30.45
7/4/2008	2	3	5.10	247.10	15.88	27.67
7/5/2008	3	5.8	7.34	212.26	15.31	23.86
7/6/2008	2.6	7.3	6.46	180.04	12.56	20.32
7/7/2008	3.1	10	7.56	154.35	10.02	17.49
7/8/2008	2.6	8.1	6.46	135.11	7.90	15.36
7/9/2008	2.5	7.1	6.23	125.76	8.08	14.32
7/10/2008	3.7	14	8.85	129.67	8.00	14.76
7/11/2008	4.8	20	11.18	124.90	7.57	14.23
7/12/2008	3.3	11	7.99	113.56	11.57	12.97
7/13/2008	2.6	8.6	6.46	117.92	13.19	13.45
7/14/2008	2.4	7.3	6.01	386.44	23.71	42.80
7/15/2008	2.6	6.8	6.46	527.09	26.64	57.92
7/16/2008	2.2	6.6	5.56	353.49	18.66	39.23
7/17/2008	2	4.7	5.10	256.44	12.83	28.69
7/18/2008	2	3.9	5.10	203.63	9.10	22.91
7/19/2008	2	3.9	5.10	174.79	11.00	19.74
7/20/2008	2	3.9	5.10	297.35	22.08	33.15
7/21/2008	2.7	5.2	6.68	1,640.83	97.95	175.26
7/22/2008	4	11	9.49	1,937.83	97.65	206.13
7/23/2008	2.6	7.5	6.46	1,382.45	66.16	148.30
7/24/2008	2.4	6.4	6.01	1,453.21	72.67	155.69
7/25/2008	2.2	5.7	5.56	1,707.88	98.92	182.24
7/26/2008	2.6	6.4	6.46	1,514.84	80.65	162.13
7/27/2008	3.2	9.5	7.77	1,275.04	96.38	137.05
7/28/2008	2.5	7.5	6.23	1,135.90	102.49	122.45
7/29/2008	2.3	6.4	5.78	910.87	67.88	98.73
7/30/2008	2.3	5.9	5.78	699.94	44.02	76.37
7/31/2008	3.3	8.3	7.99	819.19	43.19	89.03
8/1/2008	3.4	9.3	8.21	1,085.87	40.32	117.19
8/2/2008	2.8	7.5	6.90	918.69	33.07	99.56
8/3/2008	2.3	7.3	5.78	908.56	34.43	98.49
8/4/2008	2.2	6	5.56	886.69	32.67	96.18

Appendix 1-B. Measured and MOVE.1-estimated daily mean streamflow at McTier Creek near New Holland, S.C. (station 02172305) and Fishing Brook (County Line Flow) near Newcomb, N.Y. (station 0131199050).—Continued

[--, no data MOVE.1 regression technique used measured daily mean streamflow at long-term index station McTier Creek near Monetta, S.C. (station 02172300). MOVE.1 regression technique used measured daily mean streamflow at long-term index station Hudson River near Newcomb, N.Y. (station 01312000).]

| Date | Daily mean streamflow, in cubic feet per second | | | | | |
| | McTier Creek | | | Fishing Brook | | |
	Measured for MOVE.1 long-term index station 02172300	Measured for 02172305	MOVE.1 estimate for 02172305	Measured at MOVE.1 long-term index station 01312000	Measured at 0131199050	MOVE.1 estimate for 013199050
8/5/2008	2.1	5.2	5.33	754.85	27.26	82.21
8/6/2008	2.2	4.8	5.56	1,038.25	135.25	112.17
8/7/2008	2.1	5.2	5.33	1,926.55	256.94	204.96
8/8/2008	2	4.8	5.10	1,733.21	142.06	184.88
8/9/2008	2	4	5.10	1,439.88	108.41	154.30
8/10/2008	2	3.5	5.10	1,207.12	87.71	129.93
8/11/2008	2	4.4	5.10	1,096.04	85.98	118.26
8/12/2008	2	3.2	5.10	1,059.24	87.77	114.39
8/13/2008	7.9	39	17.46	938.56	68.21	101.66
8/14/2008	6.6	30	14.86	816.75	51.87	88.77
8/15/2008	5	17	11.59	647.40	39.27	70.78
8/16/2008	3.8	12	9.07	495.11	31.38	54.49
8/17/2008	3	10	7.34	392.25	25.55	43.42
8/18/2008	2.9	9.5	7.12	324.57	22.72	36.10
8/19/2008	2.7	9	6.68	431.02	50.24	47.60
8/20/2008	2.5	8.2	6.23	653.54	92.21	71.43
8/21/2008	2.5	8.6	6.23	538.78	66.11	59.17
8/22/2008	2.9	9	7.12	413.78	42.84	45.75
8/23/2008	6.2	17	14.05	325.50	30.09	36.20
8/24/2008	4.9	15	11.38	269.90	22.75	30.16
8/25/2008	3.7	12	8.85	253.95	20.56	28.42
8/26/2008	3.8	11	9.07	227.48	17.67	25.53
8/27/2008	8.9	31	19.43	194.42	15.56	21.90
8/28/2008	6.5	19	14.66	168.38	13.95	19.04
8/29/2008	5.5	15	12.63	148.52	12.22	16.85
8/30/2008	4.4	12	10.34	136.23	15.89	15.49
8/31/2008	3.4	10	8.21	129.21	45.74	14.71
9/1/2008	3	8.8	7.34	121.00	45.44	13.79
9/2/2008	2.7	8	6.68	114.07	32.93	13.02
9/3/2008	2.4	7	6.01	106.80	23.50	12.21
9/4/2008	2.3	6.3	5.78	99.95	15.81	11.45
9/5/2008	2.3	6	5.78	92.91	12.40	10.66
9/6/2008	2.4	6.5	6.01	89.35	12.28	10.26
9/7/2008	2.4	6	6.01	93.92	15.34	10.78
9/8/2008	2.9	7.7	7.12	87.14	16.01	10.02
9/9/2008	2.9	7.3	7.12	91.27	16.55	10.48

Appendix 1-B.   Measured and MOVE.1-estimated daily mean streamflow at McTier Creek near New Holland, S.C. (station 02172305) and Fishing Brook (County Line Flow) near Newcomb, N.Y. (station 0131199050).—Continued

[--, no data MOVE.1 regression technique used measured daily mean streamflow at long-term index station McTier Creek near Monetta, S.C. (station 02172300). MOVE.1 regression technique used measured daily mean streamflow at long-term index station Hudson River near Newcomb, N.Y. (station 01312000).]

	Daily mean streamflow, in cubic feet per second					
	McTier Creek			Fishing Brook		
Date	Measured for MOVE.1 long-term index station 02172300	Measured for 02172305	MOVE.1 estimate for 02172305	Measured at MOVE.1 long-term index station 01312000	Measured at 0131199050	MOVE.1 estimate for 013199050
9/10/2008	6.7	15	15.07	99.86	19.44	11.44
9/11/2008	5.1	15	11.80	93.09	18.35	10.68
9/12/2008	5.1	17	11.80	87.87	16.19	10.10
9/13/2008	4.8	14	11.18	106.94	20.50	12.23
9/14/2008	4	11	9.49	197.42	37.76	22.23
9/15/2008	3.6	9.7	8.64	472.66	60.44	52.08
9/16/2008	3.4	9.1	8.21	410.10	46.32	45.35
9/17/2008	3.6	9.5	8.64	306.92	31.73	34.19
9/18/2008	3.6	9.6	8.64	244.26	22.84	27.36
9/19/2008	3.6	8.6	8.64	197.04	16.80	22.19
9/20/2008	3.1	8.1	7.56	165.04	13.33	18.67
9/21/2008	3	7.6	7.34	143.68	10.09	16.31
9/22/2008	3.1	7.8	7.56	128.06	8.79	14.58
9/23/2008	2.7	7	6.68	114.52	8.23	13.07
9/24/2008	2.5	6.4	6.23	103.84	7.23	11.88
9/25/2008	2.3	5.7	5.78	94.40	6.33	10.83
9/26/2008	3	7.8	7.34	89.19	6.47	10.25
9/27/2008	3.7	12	8.85	94.24	8.56	10.81
9/28/2008	3.8	11	9.07	97.51	9.97	11.18
9/29/2008	3.8	9.8	9.07	95.06	12.34	10.90
9/30/2008	5.4	20	12.42	93.47	14.51	10.73
10/1/2008	4.4	16	10.34	106.20	14.85	12.15
10/2/2008	3.7	11	8.85	150.00	16.50	17.01
10/3/2008	3.5	9	8.42	185.69	23.98	20.94
10/4/2008	3.3	8.1	7.99	212.81	34.04	23.92
10/5/2008	3.2	7.8	7.77	208.16	34.38	23.41
10/6/2008	3.2	7.4	7.77	188.81	27.59	21.29
10/7/2008	2.9	7.4	7.12	168.91	21.88	19.10
10/8/2008	3.3	10	7.99	152.09	18.25	17.24
10/9/2008	14	72	29.14	171.87	20.44	19.42
10/10/2008	7.7	34	17.06	278.31	21.31	31.08
10/11/2008	6.8	24	15.27	264.19	21.41	29.54
10/12/2008	5.3	19	12.21	220.60	19.73	24.77
10/13/2008	4.4	16	10.34	187.81	17.69	21.18
10/14/2008	4.1	14	9.71	163.02	16.84	18.45
10/15/2008	3.9	13	9.28	144.49	14.76	16.40

Appendix 1-B. Measured and MOVE.1-estimated daily mean streamflow at McTier Creek near New Holland, S.C. (station 02172305) and Fishing Brook (County Line Flow) near Newcomb, N.Y. (station 0131199050).—Continued

[--, no data MOVE.1 regression technique used measured daily mean streamflow at long-term index station McTier Creek near Monetta, S.C. (station 02172300). MOVE.1 regression technique used measured daily mean streamflow at long-term index station Hudson River near Newcomb, N.Y. (station 01312000).]

Date	Daily mean streamflow, in cubic feet per second					
	McTier Creek			Fishing Brook		
	Measured for MOVE.1 long-term index station 02172300	Measured for 02172305	MOVE.1 estimate for 02172305	Measured at MOVE.1 long-term index station 01312000	Measured at 0131199050	MOVE.1 estimate for 013199050
10/16/2008	3.7	12	8.85	141.88	19.39	16.11
10/17/2008	3.8	12	9.07	164.09	28.68	18.57
10/18/2008	4.8	16	11.18	168.85	32.17	19.09
10/19/2008	4.1	15	9.71	156.68	27.77	17.75
10/20/2008	3.9	13	9.28	144.82	23.65	16.44
10/21/2008	4	13	9.49	142.56	23.41	16.19
10/22/2008	3.9	12	9.28	156.26	24.53	17.70
10/23/2008	3.8	12	9.07	157.04	24.94	17.79
10/24/2008	6	21	13.65	146.61	23.80	16.63
10/25/2008	9.8	48	21.18	157.64	39.23	17.85
10/26/2008	7.7	28	17.06	1,029.47	268.84	111.25
10/27/2008	5.7	20	13.04	1,821.96	219.61	194.10
10/28/2008	5.1	16	11.80	1,437.10	129.22	154.01
10/29/2008	4.9	17	11.38	1,136.12	118.59	122.47
10/30/2008	4.6	15	10.76	936.34	96.21	101.43
10/31/2008	4.4	14	10.34	771.43	77.40	83.97
11/1/2008	4.4	13	10.34	677.56	82.73	73.99
11/2/2008	4.3	13	10.13	704.96	94.79	76.91
11/3/2008	4.2	13	9.92	644.95	83.39	70.52
11/4/2008	4.4	14	10.34	557.87	71.06	61.22
11/5/2008	4.8	16	11.18	516.29	81.76	56.76
11/6/2008	4.9	16	11.38	564.37	97.45	61.91
11/7/2008	5.1	15	11.80	580.68	86.55	63.66
11/8/2008	4.7	14	10.97	552.88	73.30	60.68
11/9/2008	4.5	14	10.55	546.39	67.56	59.99
11/10/2008	4.6	14	10.76	537.10	64.86	58.99
11/11/2008	4.5	15	10.55	486.43	58.62	53.56
11/12/2008	4.8	15	11.18	422.75	50.94	46.71
11/13/2008	7.4	25	16.47	369.06	46.52	40.92
11/14/2008	93	113	158.78	357.90	53.02	39.71
11/15/2008	55	154	99.21	506.73	74.75	55.74
11/16/2008	18	49	36.50	1,068.92	149.92	115.40
11/17/2008	13	31	27.27	1,416.01	145.40	151.81
11/18/2008	12	23	25.39	1,177.39	100.77	126.81
11/19/2008	11	20	23.48	917.54	73.24	99.44
11/20/2008	11	21	23.48	712.45	57.08	77.70

Appendix 1-B.   Measured and MOVE.1-estimated daily mean streamflow at McTier Creek near New Holland, S.C. (station 02172305) and Fishing Brook (County Line Flow) near Newcomb, N.Y. (station 0131199050).—Continued

[--, no data MOVE.1 regression technique used measured daily mean streamflow at long-term index station McTier Creek near Monetta, S.C. (station 02172300). MOVE.1 regression technique used measured daily mean streamflow at long-term index station Hudson River near Newcomb, N.Y. (station 01312000).]

	Daily mean streamflow, in cubic feet per second					
	McTier Creek			Fishing Brook		
Date	Measured for MOVE.1 long-term index station 02172300	Measured for 02172305	MOVE.1 estimate for 02172305	Measured at MOVE.1 long-term index station 01312000	Measured at 0131199050	MOVE.1 estimate for 013199050
11/21/2008	10	21	21.56	532.10	47.67	58.46
11/22/2008	9.9	16	21.37	411.21	41.54	45.47
11/23/2008	10	15	21.56	332.78	36.87	36.99
11/24/2008	10	16	21.56	295.56	34.48	32.95
11/25/2008	10	17	21.56	284.51	37.79	31.75
11/26/2008	10	16	21.56	285.70	39.33	31.88
11/27/2008	10	16	21.56	272.33	36.88	30.42
11/28/2008	9.6	16	20.79	253.81	34.98	28.41
11/29/2008	24	42	47.22	239.49	33.93	26.84
11/30/2008	92	122	157.25	223.18	32.53	25.06
12/1/2008	44	131	81.24	246.85	43.40	27.65
12/2/2008	19	56	38.31	292.67	56.07	32.64
12/3/2008	15	35	31.00	293.26	52.65	32.70
12/4/2008	13	30	27.27	274.76	45.24	30.69
12/5/2008	13	29	27.27	256.35	41.66	28.68
12/6/2008	12	26	25.39	225.79	36.56	25.34
12/7/2008	12	24	25.39	219.54	34.69	24.66
12/8/2008	12	22	25.39	190.67	29.25	21.49
12/9/2008	11	22	23.48	181.09	27.18	20.44
12/10/2008	15	26	31.00	224.37	56.73	25.19
12/11/2008	103	125	173.98	468.32	106.37	51.62
12/12/2008	123	163	203.94	685.22	103.01	74.81
12/13/2008	40	115	74.60	547.66	76.40	60.12
12/14/2008	23	54	45.45	412.82	58.07	45.64
12/15/2008	19	43	38.31	382.23	59.72	42.34
12/16/2008	17	38	34.68	468.89	107.83	51.68
12/17/2008	16	35	32.84	640.31	109.52	70.02
12/18/2008	15	33	31.00	585.13	85.56	64.13
12/19/2008	14	31	29.14	495.21	68.02	54.50
12/20/2008	15	31	31.00	407.09	57.71	45.02
12/21/2008	14	30	29.14	368.28	51.41	40.83
12/22/2008	13	27	27.27	332.25	42.99	36.93
12/23/2008	12	27	25.39	305.77	42.95	34.06
12/24/2008	13	30	27.27	294.33	40.88	32.82
12/25/2008	13	27	27.27	322.42	58.93	35.87
12/26/2008	14	28	29.14	416.06	90.78	45.99

Appendix 1-B.   Measured and MOVE.1-estimated daily mean streamflow at McTier Creek near New Holland, S.C. (station 02172305) and Fishing Brook (County Line Flow) near Newcomb, N.Y. (station 0131199050).—Continued

[--, no data MOVE.1 regression technique used measured daily mean streamflow at long-term index station McTier Creek near Monetta, S.C. (station 02172300). MOVE.1 regression technique used measured daily mean streamflow at long-term index station Hudson River near Newcomb, N.Y. (station 01312000).]

| Date | Daily mean streamflow, in cubic feet per second | | | | | |
| | McTier Creek | | | Fishing Brook | | |
	Measured for MOVE.1 long-term index station 02172300	Measured for 02172305	MOVE.1 estimate for 02172305	Measured at MOVE.1 long-term index station 01312000	Measured at 0131199050	MOVE.1 estimate for 013199050
12/27/2008	14	29	29.14	478.75	81.95	52.74
12/28/2008	14	28	29.14	671.47	148.18	73.34
12/29/2008	13	27	27.27	1,201.76	260.06	129.37
12/30/2008	14	27	29.14	1,338.80	180.35	143.73
12/31/2008	13	25	27.27	1,119.58	115.30	120.73
1/1/2009	11	22	23.48	884.06	82.33	95.90
1/2/2009	12	26	25.39	688.73	64.39	75.18
1/3/2009	12	25	25.39	560.26	59.19	61.47
1/4/2009	12	25	25.39	489.36	50.59	53.88
1/5/2009	13	26	27.27	411.25	47.13	45.47
1/6/2009	13	27	27.27	354.95	43.24	39.39
1/7/2009	16	36	32.84	326.18	44.04	36.28
1/8/2009	15	34	31.00	310.28	41.08	34.55
1/9/2009	12	26	25.39	286.81	38.25	32.00
1/10/2009	12	24	25.39	271.24	36.29	30.31
1/11/2009	15	34	31.00	257.16	36.92	28.77
1/12/2009	14	31	29.14	241.10	34.34	27.02
1/13/2009	13	28	27.27	226.35	33.04	25.40
1/14/2009	12	29	25.39	215.32	31.10	24.20
1/15/2009	12	28	25.39	203.93	29.71	22.95
1/16/2009	11	28	23.48	195.21	27.89	21.99
1/17/2009	11	24	23.48	187.50	26.48	21.14
1/18/2009	13	26	27.27	181.91	26.27	20.53
1/19/2009	14	33	29.14	177.36	26.44	20.03
1/20/2009	13	28	27.27	171.53	26.37	19.39
1/21/2009	12	27	25.39	164.80	25.60	18.64
1/22/2009	12	31	25.39	160.93	24.84	18.22
1/23/2009	11	26	23.48	156.92	24.07	17.77
1/24/2009	12	25	25.39	152.80	24.27	17.32
1/25/2009	12	25	25.39	147.30	23.61	16.71
1/26/2009	12	23	25.39	144.72	22.79	16.43
1/27/2009	12	25	25.39	141.10	22.09	16.03
1/28/2009	12	26	25.39	140.67	22.98	15.98
1/29/2009	12	26	25.39	149.07	23.53	16.91
1/30/2009	12	23	25.39	149.31	23.55	16.93
1/31/2009	11	21	23.48	146.90	22.69	16.67

Appendix 1-B. Measured and MOVE.1-estimated daily mean streamflow at McTier Creek near New Holland, S.C. (station 02172305) and Fishing Brook (County Line Flow) near Newcomb, N.Y. (station 0131199050).—Continued

[--, no data MOVE.1 regression technique used measured daily mean streamflow at long-term index station McTier Creek near Monetta, S.C. (station 02172300). MOVE.1 regression technique used measured daily mean streamflow at long-term index station Hudson River near Newcomb, N.Y. (station 01312000).]

| | Daily mean streamflow, in cubic feet per second | | | | | |
| | McTier Creek | | | Fishing Brook | | |
Date	Measured for MOVE.1 long-term index station 02172300	Measured for 02172305	MOVE.1 estimate for 02172305	Measured at MOVE.1 long-term index station 01312000	Measured at 0131199050	MOVE.1 estimate for 013199050
2/1/2009	11	24	23.48	142.94	22.06	16.23
2/2/2009	11	26	23.48	137.90	21.41	15.67
2/3/2009	12	25	25.39	132.99	21.00	15.13
2/4/2009	11	22	23.48	128.41	20.24	14.62
2/5/2009	11	20	23.48	124.91	19.19	14.23
2/6/2009	11	20	23.48	122.25	18.94	13.93
2/7/2009	11	21	23.48	120.27	19.04	13.71
2/8/2009	11	20	23.48	124.80	23.04	14.22
2/9/2009	11	20	23.48	128.22	27.62	14.60
2/10/2009	11	20	23.48	128.87	24.90	14.67
2/11/2009	11	20	23.48	128.40	24.52	14.62
2/12/2009	11	20	23.48	150.12	47.56	17.02
2/13/2009	11	19	23.48	205.48	74.66	23.12
2/14/2009	13	26	27.27	291.52	68.87	32.51
2/15/2009	13	30	27.27	333.94	50.30	37.12
2/16/2009	12	23	25.39	308.90	39.73	34.40
2/17/2009	11	21	23.48	268.72	33.80	30.03
2/18/2009	14	28	29.14	235.88	30.83	26.45
2/19/2009	65	121	115.21	218.12	30.01	24.50
2/20/2009	21	64	41.90	200.85	27.69	22.61
2/21/2009	15	39	31.00	185.56	26.28	20.93
2/22/2009	14	33	29.14	177.95	27.23	20.09
2/23/2009	13	27	27.27	170.94	25.54	19.32
2/24/2009	12	24	25.39	161.51	23.55	18.28
2/25/2009	12	23	25.39	152.02	23.31	17.23
2/26/2009	12	22	25.39	144.92	22.81	16.45
2/27/2009	14	25	29.14	147.99	31.94	16.79
2/28/2009	95	97	161.83	199.79	78.51	22.49
3/1/2009	116	159	193.51	303.70	92.11	33.84
3/2/2009	54	130	97.59	401.26	69.98	44.40
3/3/2009	30	64	57.66	366.52	48.25	40.64
3/4/2009	24	47	47.22	305.06	37.40	33.98
3/5/2009	21	41	41.90	260.01	32.35	29.08
3/6/2009	19	38	38.31	230.21	30.79	25.83
3/7/2009	18	35	36.50	214.70	38.96	24.13
3/8/2009	16	33	32.84	266.60	77.85	29.80

Appendix 1-B. Measured and MOVE.1-estimated daily mean streamflow at McTier Creek near New Holland, S.C. (station 02172305) and Fishing Brook (County Line Flow) near Newcomb, N.Y. (station 0131199050).—Continued

[--, no data MOVE.1 regression technique used measured daily mean streamflow at long-term index station McTier Creek near Monetta, S.C. (station 02172300). MOVE.1 regression technique used measured daily mean streamflow at long-term index station Hudson River near Newcomb, N.Y. (station 01312000).]

	Daily mean streamflow, in cubic feet per second					
	McTier Creek			Fishing Brook		
Date	**Measured for MOVE.1 long-term index station 02172300**	**Measured for 02172305**	**MOVE.1 estimate for 02172305**	**Measured at MOVE.1 long-term index station 01312000**	**Measured at 0131199050**	**MOVE.1 estimate for 013199050**
3/9/2009	16	31	32.84	429.96	112.81	47.49
3/10/2009	15	29	31.00	594.54	100.75	65.14
3/11/2009	15	29	31.00	642.74	94.35	70.28
3/12/2009	15	28	31.00	739.80	119.03	80.61
3/13/2009	14	27	29.14	792.68	108.75	86.22
3/14/2009	14	27	29.14	719.67	82.30	78.47
3/15/2009	20	38	40.11	580.78	65.73	63.67
3/16/2009	27	55	52.47	485.98	60.47	53.51
3/17/2009	22	50	43.68	450.34	66.16	49.68
3/18/2009	17	36	34.68	499.63	86.53	54.98
3/19/2009	15	31	31.00	630.02	118.61	68.92
3/20/2009	14	27	29.14	775.48	123.90	84.40
3/21/2009	14	25	29.14	787.73	100.15	85.70
3/22/2009	14	26	29.14	721.11	85.11	78.62
3/23/2009	13	24	27.27	579.80	64.26	63.56
3/24/2009	13	24	27.27	499.11	55.73	54.92
3/25/2009	14	25	29.14	433.68	54.18	47.89
3/26/2009	14	25	29.14	402.01	56.39	44.48
3/27/2009	15	30	31.00	467.16	81.48	51.49
3/28/2009	49	59	89.46	753.16	121.12	82.03
3/29/2009	69	134	121.54	1,130.48	189.64	121.88
3/30/2009	24	60	47.22	1,873.77	352.97	199.48
3/31/2009	18	39	36.50	2,144.37	237.33	227.52
4/1/2009	16	33	32.84	1,747.97	148.41	186.41
4/2/2009	32	57	61.09	1,406.03	123.47	150.76
4/3/2009	41	82	76.27	1,436.38	161.15	153.93
4/4/2009	24	52	47.22	2,309.60	310.06	244.60
4/5/2009	18	38	36.50	2,675.45	233.32	282.30
4/6/2009	17	35	34.68	2,009.32	145.77	213.54
4/7/2009	16	32	32.84	1,558.76	126.42	166.71
4/8/2009	15	30	31.00	1,328.40	111.63	142.64
4/9/2009	14	27	29.14	1,103.45	95.08	119.04
4/10/2009	14	26	29.14	951.57	83.96	103.03
4/11/2009	86	105	148.03	879.04	79.89	95.37
4/12/2009	31	81	59.38	848.72	75.68	92.16
4/13/2009	22	44	43.68	772.83	67.98	84.12

Appendix 1-B.   Measured and MOVE.1-estimated daily mean streamflow at McTier Creek near New Holland, S.C. (station 02172305) and Fishing Brook (County Line Flow) near Newcomb, N.Y. (station 0131199050).—Continued

[--, no data MOVE.1 regression technique used measured daily mean streamflow at long-term index station McTier Creek near Monetta, S.C. (station 02172300). MOVE.1 regression technique used measured daily mean streamflow at long-term index station Hudson River near Newcomb, N.Y. (station 01312000).]

| Date | Daily mean streamflow, in cubic feet per second | | | | | |
| | McTier Creek | | | Fishing Brook | | |
	Measured for MOVE.1 long-term index station 02172300	Measured for 02172305	MOVE.1 estimate for 02172305	Measured at MOVE.1 long-term index station 01312000	Measured at 0131199050	MOVE.1 estimate for 013199050
4/14/2009	31	55	59.38	676.22	62.42	73.85
4/15/2009	32	73	61.09	632.38	60.51	69.18
4/16/2009	21	45	41.90	672.77	61.77	73.48
4/17/2009	18	36	36.50	740.70	64.96	80.70
4/18/2009	17	32	34.68	832.19	71.49	90.41
4/19/2009	17	29	34.68	936.00	91.60	101.39
4/20/2009	18	37	36.50	957.74	92.05	103.69
4/21/2009	17	33	34.68	947.48	100.66	102.60
4/22/2009	15	28	31.00	1,031.88	122.97	111.50
4/23/2009	14	25	29.14	1,177.28	123.07	126.80
4/24/2009	14	23	29.14	1,118.64	104.95	120.64
4/25/2009	13	22	27.27	1,089.00	103.62	117.52
4/26/2009	12	20	25.39	1,466.19	142.61	157.05
4/27/2009	13	20	27.27	1,724.70	132.32	183.99
4/28/2009	12	19	25.39	1,668.47	113.83	178.14
4/29/2009	12	19	25.39	1,604.90	98.34	171.52
4/30/2009	12	19	25.39	1,299.97	77.80	139.66
5/1/2009	11	18	23.48	1,080.44	79.41	116.62
5/2/2009	12	18	25.39	1,437.44	105.14	154.05
5/3/2009	12	20	25.39	1,410.77	98.90	151.26
5/4/2009	11	19	23.48	1,100.31	76.78	118.71
5/5/2009	15	33	31.00	880.06	60.39	95.48
5/6/2009	15	35	31.00	786.79	64.54	85.60
5/7/2009	15	34	31.00	850.45	91.95	92.34
5/8/2009	13	30	27.27	1,074.55	108.81	116.00
5/9/2009	12	22	25.39	1,100.87	104.38	118.77
5/10/2009	11	20	23.48	1,378.93	172.52	147.93
5/11/2009	11	20	23.48	1,534.72	165.14	164.20
5/12/2009	11	21	23.48	1,263.03	121.60	135.79
5/13/2009	10	18	21.56	1,016.79	92.74	109.91
5/14/2009	9.5	17	20.59	853.25	75.79	92.64
5/15/2009	9.5	17	20.59	854.46	76.47	92.77
5/16/2009	12	28	25.39	860.14	81.45	93.37
5/17/2009	12	46	25.39	1,623.80	235.31	173.49
5/18/2009	14	55	29.14	2,133.15	202.91	226.36
5/19/2009	11	32	23.48	1,606.50	126.55	171.69
5/20/2009	10	23	21.56	1,156.23	92.72	124.59

Appendix 1-B. Measured and MOVE.1-estimated daily mean streamflow at McTier Creek near New Holland, S.C. (station 02172305) and Fishing Brook (County Line Flow) near Newcomb, N.Y. (station 0131199050).—Continued

[--, no data MOVE.1 regression technique used measured daily mean streamflow at long-term index station McTier Creek near Monetta, S.C. (station 02172300). MOVE.1 regression technique used measured daily mean streamflow at long-term index station Hudson River near Newcomb, N.Y. (station 01312000).]

Date	Daily mean streamflow, in cubic feet per second					
	McTier Creek			Fishing Brook		
	Measured for MOVE.1 long-term index station 02172300	Measured for 02172305	MOVE.1 estimate for 02172305	Measured at MOVE.1 long-term index station 01312000	Measured at 0131199050	MOVE.1 estimate for 013199050
5/21/2009	10	25	21.56	914.90	72.13	99.16
5/22/2009	10	22	21.56	744.22	58.68	81.08
5/23/2009	8.5	21	18.64	576.94	48.11	63.26
5/24/2009	10	23	21.56	464.59	44.45	51.21
5/25/2009	12	30	25.39	388.14	42.18	42.98
5/26/2009	11	26	23.48	318.49	35.05	35.44
5/27/2009	9.1	23	19.82	318.96	39.38	35.49
5/28/2009	11	31	23.48	797.27	90.73	86.71
5/29/2009	13	30	27.27	1,645.15	127.04	175.71
5/30/2009	10	21	21.56	1,835.88	112.20	195.55
5/31/2009	11	20	23.48	1,499.48	89.28	160.53
6/1/2009	8.5	16	18.64	1,113.12	65.68	120.05
6/2/2009	6.8	13	15.27	876.46	53.40	95.10
6/3/2009	6	11	13.65	691.00	44.61	75.42
6/4/2009	5.7	17	13.04	520.46	38.84	57.21
6/5/2009	16	51	32.84	405.18	37.78	44.82
6/6/2009	12	34	25.39	330.19	33.48	36.71
6/7/2009	9.3	23	20.21	280.92	27.57	31.36
6/8/2009	8.1	20	17.86	246.54	24.34	27.61
6/9/2009	7.2	17	16.07	246.30	25.26	27.59
6/10/2009	6.3	14	14.26	264.84	26.24	29.61
6/11/2009	6.1	13	13.85	245.17	23.74	27.46
6/12/2009	5.8	15	13.24	315.08	43.50	35.07
6/13/2009	5.6	13	12.83	460.95	54.45	50.82
6/14/2009	5.3	12	12.21	410.45	48.02	45.39
6/15/2009	5	11	11.59	370.87	52.97	41.11
6/16/2009	4.8	10	11.18	585.13	51.19	64.13
6/17/2009	5.4	11	12.42	602.51	39.52	65.99
6/18/2009	5.6	12	12.83	458.46	32.65	50.56
6/19/2009	5	11	11.59	637.64	39.48	69.74
6/20/2009	4.6	9.8	10.76	805.20	42.13	87.55
6/21/2009	4.6	8.9	10.76	650.04	37.34	71.06
6/22/2009	5.3	9.5	12.21	498.38	31.22	54.84
6/23/2009	6.3	12	14.26	424.23	25.82	46.87
6/24/2009	4.8	9.2	11.18	330.81	20.91	36.78
6/25/2009	4.2	7.6	9.92	267.42	17.03	29.89
6/26/2009	3.8	7	9.07	224.40	14.68	25.19

Appendix 1-B.   Measured and MOVE.1-estimated daily mean streamflow at McTier Creek near New Holland, S.C. (station 02172305) and Fishing Brook (County Line Flow) near Newcomb, N.Y. (station 0131199050).—Continued

[--, no data MOVE.1 regression technique used measured daily mean streamflow at long-term index station McTier Creek near Monetta, S.C. (station 02172300). MOVE.1 regression technique used measured daily mean streamflow at long-term index station Hudson River near Newcomb, N.Y. (station 01312000).]

| Date | Daily mean streamflow, in cubic feet per second | | | | | |
| | McTier Creek | | | Fishing Brook | | |
	Measured for MOVE.1 long-term index station 02172300	Measured for 02172305	MOVE.1 estimate for 02172305	Measured at MOVE.1 long-term index station 01312000	Measured at 0131199050	MOVE.1 estimate for 013199050
6/27/2009	3.5	6.7	8.42	215.44	18.87	24.21
6/28/2009	3.6	6	8.64	239.78	30.02	26.87
6/29/2009	3.4	5.6	8.21	247.28	32.99	27.69
6/30/2009	3.2	5.1	7.77	235.26	27.81	26.38
7/1/2009	2.9	4.5	7.12	241.26	25.44	27.03
7/2/2009	2.7	3.9	6.68	249.87	21.94	27.98
7/3/2009	2.6	3.7	6.46	240.10	23.14	26.91
7/4/2009	2.6	3.8	6.46	337.96	36.45	37.55
7/5/2009	2.7	4.4	6.68	758.47	55.17	82.59
7/6/2009	4.2	9.3	9.92	721.80	65.73	78.70
7/7/2009	3.7	8.4	8.85	528.19	53.57	58.04
7/8/2009	7.5	44	16.67	566.53	53.11	62.14
7/9/2009	10	69	21.56	703.07	54.06	76.70
7/10/2009	6.4	25	14.46	621.30	44.44	67.99
7/11/2009	5.3	16	12.21	492.14	36.63	54.17
7/12/2009	4.4	12	10.34	615.34	53.98	67.36
7/13/2009	4.8	14	11.18	761.74	53.48	82.94
7/14/2009	4.6	15	10.76	617.01	42.08	67.54
7/15/2009	4	12	9.49	495.04	33.84	54.48
7/16/2009	4.1	11	9.71	409.36	26.92	45.27
7/17/2009	3.7	10	8.85	344.45	21.03	38.26
7/18/2009	3.7	11	8.85	315.18	17.87	35.08
7/19/2009	3.2	8.7	7.77	290.82	14.62	32.44
7/20/2009	3.1	7.9	7.56	255.67	12.78	28.61
7/21/2009	3	7.7	7.34	223.70	11.29	25.11
7/22/2009	3	7.6	7.34	204.99	10.82	23.06
7/23/2009	3.2	8.7	7.77	186.84	9.99	21.07
7/24/2009	3	7.6	7.34	171.10	10.38	19.34
7/25/2009	2.7	7.9	6.68	166.68	11.25	18.85
7/26/2009	2.4	6.7	6.01	159.23	13.28	18.03
7/27/2009	2.4	5.8	6.01	149.62	14.19	16.97
7/28/2009	2.4	5.7	6.01	138.16	14.16	15.70
7/29/2009	2.4	6.2	6.01	130.26	14.13	14.82
7/30/2009	3.3	9.1	7.99	162.70	26.29	18.41
7/31/2009	5.7	20	13.04	214.97	41.03	24.16
8/1/2009	6.3	23	14.26	327.17	50.04	36.38
8/2/2009	5.1	16	11.80	340.78	50.31	37.86

Appendix 1-B.   Measured and MOVE.1-estimated daily mean streamflow at McTier Creek near New Holland, S.C. (station 02172305) and Fishing Brook (County Line Flow) near Newcomb, N.Y. (station 0131199050).—Continued

[--, no data MOVE 1 regression technique used measured daily mean streamflow at long-term index station McTier Creek near Monetta, S.C. (station 02172300). MOVE 1 regression technique used measured daily mean streamflow at long-term index station Hudson River near Newcomb, N.Y. (station 01312000).]

	Daily mean streamflow, in cubic feet per second					
	McTier Creek			Fishing Brook		
Date	Measured for MOVE.1 long-term index station 02172300	Measured for 02172305	MOVE.1 estimate for 02172305	Measured at MOVE.1 long-term index station 01312000	Measured at 0131199050	MOVE.1 estimate for 013199050
8/3/2009	4.8	15	11.18	463.29	74.75	51.07
8/4/2009	3.8	12	9.07	507.23	57.04	55.79
8/5/2009	3.5	10	8.42	406.81	37.46	44.99
8/6/2009	3.2	8.8	7.77	320.30	25.90	35.64
8/7/2009	2.8	7.8	6.90	259.35	18.61	29.01
8/8/2009	2.7	7	6.68	213.59	14.15	24.01
8/9/2009	2.6	6.8	6.46	180.88	11.68	20.42
8/10/2009	2.4	7.2	6.01	162.19	11.17	18.36
8/11/2009	2.4	5.4	6.01	155.10	11.08	17.57
8/12/2009	3.9	9.3	9.28	143.70	9.57	16.31
8/13/2009	32	72	61.09	194.29	8.66	21.89
8/14/2009	23	86	45.45	223.83	8.50	25.13
8/15/2009	10	36	21.56	201.54	7.69	22.69
8/16/2009	8.4	21	18.45	160.37	7.19	18.15
8/17/2009	8.8	21	19.23	132.23	6.54	15.04
8/18/2009	6.7	14	15.07	115.01	6.52	13.13
8/19/2009	5.6	11	12.83	117.54	11.82	13.41
8/20/2009	5.8	12	13.24	126.83	15.78	14.44
8/21/2009	7.1	16	15.87	188.73	18.35	21.28
8/22/2009	5.8	11	13.24	181.19	19.50	20.45
8/23/2009	4.8	8.6	11.18	172.76	33.92	19.52
8/24/2009	4.4	7.3	10.34	199.25	70.83	22.43
8/25/2009	4.2	8	9.92	236.52	52.61	26.52
8/26/2009	4.2	7.1	9.92	216.06	33.37	24.28
8/27/2009	4.3	7	10.13	188.04	23.92	21.20
8/28/2009	4.8	8.2	11.18	161.42	17.95	18.27
8/29/2009	4.8	8.8	11.18	193.92	22.92	21.85
8/30/2009	6.5	15	14.66	344.77	31.70	38.29
8/31/2009	12	44	25.39	387.16	31.16	42.87
9/1/2009	10	24	21.56	302.77	24.11	33.74
9/2/2009	9.9	19	21.37	239.41	18.74	26.83
9/3/2009	7.6	14	16.87	196.29	14.87	22.11
9/4/2009	6	11	13.65	165.38	12.20	18.71
9/5/2009	5.2	8.9	12.01	140.84	10.25	16.00
9/6/2009	4.8	8.2	11.18	122.63	7.98	13.98
9/7/2009	4.5	7.3	10.55	107.97	7.09	12.34
9/8/2009	4.5	7.2	10.55	97.37	6.61	11.16

Appendix 1-B.   Measured and MOVE.1-estimated daily mean streamflow at McTier Creek near New Holland, S.C. (station 02172305) and Fishing Brook (County Line Flow) near Newcomb, N.Y. (station 0131199050).—Continued

[--, no data MOVE.1 regression technique used measured daily mean streamflow at long-term index station McTier Creek near Monetta, S.C. (station 02172300). MOVE.1 regression technique used measured daily mean streamflow at long-term index station Hudson River near Newcomb, N.Y. (station 01312000).]

| Date | Daily mean streamflow, in cubic feet per second | | | | | |
| | McTier Creek | | | Fishing Brook | | |
	Measured for MOVE.1 long-term index station 02172300	Measured for 02172305	MOVE.1 estimate for 02172305	Measured at MOVE.1 long-term index station 01312000	Measured at 0131199050	MOVE.1 estimate for 013199050
9/9/2009	4.6	7.3	10.76	88.81	5.87	10.20
9/10/2009	6.2	8.6	14.05	80.67	5.45	9.29
9/11/2009	6.1	10	13.85	71.75	5.16	8.29
9/12/2009	5.9	9.1	13.44	66.61	5.56	7.71
9/13/2009	5.5	8.8	12.63	65.03	6.27	7.53
9/14/2009	5.1	7.9	11.80	61.71	6.26	7.15
9/15/2009	4.9	7.6	11.38	61.75	6.49	7.16
9/16/2009	4.9	7.6	11.38	59.13	10.21	6.86
9/17/2009	4.8	7.1	11.18	55.49	12.89	6.45
9/18/2009	6.7	12	15.07	53.53	12.66	6.23
9/19/2009	7.9	14	17.46	53.54	9.68	6.23
9/20/2009	7.5	13	16.67	51.71	7.67	6.02
9/21/2009	10	21	21.56	48.54	6.55	5.66
9/22/2009	7.5	14	16.67	48.68	7.19	5.68
9/23/2009	6.6	11	14.86	52.11	8.90	6.07
9/24/2009	6.5	11	14.66	62.77	11.23	7.27
9/25/2009	6.1	10	13.85	61.17	11.49	7.09
9/26/2009	6	9.8	13.65	57.34	10.36	6.66
9/27/2009	8.2	17	18.05	71.62	15.33	8.27
9/28/2009	7.3	13	16.27	282.32	42.53	31.51
9/29/2009	6.2	10	14.05	450.05	63.42	49.65
9/30/2009	5.6	9	12.83	507.22	56.92	55.79

Appendix 1-C. Daily mean streamflow for selected sites from October 1, 2004, to October 1, 2009, used as input into the load estimation model at each site.

[ft³/s, cubic feet per second; Bolded values were estimated using MOVE.1 regression technique]

Date	Time	Daily mean streamflow (ft³/s)				Date	Time	Daily mean streamflow (ft³/s)			
		McTier Creek 02172305	Fishing Brook 0131199050	Edisto River 02175000	Hudson River 01312000			McTier Creek 02172305	Fishing Brook 0131199050	Edisto River 02175000	Hudson River 01312000
10/1/2004	1200	**25.39**	**12.41**	1,436.9	108.5	11/10/2004	1200	**25.39**	**34.96**	690.4	314.0
10/2/2004	1200	**25.39**	**12.17**	1,500.0	106.4	11/11/2004	1200	**25.39**	**31.00**	700.6	277.7
10/3/2004	1200	**25.39**	**12.81**	1,590.0	112.1	11/12/2004	1200	**45.45**	**27.66**	710.3	247.0
10/4/2004	1200	**25.39**	**13.38**	1,692.8	117.2	11/13/2004	1200	**54.21**	**24.39**	709.2	217.1
10/5/2004	1200	**23.48**	**13.01**	1,724.4	113.9	11/14/2004	1200	**34.68**	**21.94**	702.5	194.8
10/6/2004	1200	**23.48**	**12.37**	1,651.1	108.2	11/15/2004	1200	**29.14**	**20.59**	692.0	182.5
10/7/2004	1200	**23.48**	**11.87**	1,569.8	103.8	11/16/2004	1200	**25.39**	**19.27**	683.2	170.4
10/8/2004	1200	**23.48**	**11.46**	1,607.8	100.0	11/17/2004	1200	**27.27**	**18.36**	682.0	162.3
10/9/2004	1200	**23.48**	**11.03**	1,755.9	96.2	11/18/2004	1200	**29.14**	**17.76**	682.5	156.8
10/10/2004	1200	**23.48**	**10.66**	1,848.4	92.9	11/19/2004	1200	**27.27**	**17.62**	682.2	155.5
10/11/2004	1200	**23.48**	**10.36**	1,815.9	90.2	11/20/2004	1200	**27.27**	**17.44**	679.4	153.9
10/12/2004	1200	**21.56**	**9.94**	1,690.0	86.5	11/21/2004	1200	**25.39**	**18.42**	678.7	162.8
10/13/2004	1200	**23.48**	**9.69**	1,500.0	84.3	11/22/2004	1200	**27.27**	**20.01**	685.2	177.2
10/14/2004	1200	**23.48**	**9.36**	1,299.3	81.3	11/23/2004	1200	**41.90**	**21.34**	699.1	189.3
10/15/2004	1200	**25.39**	**9.18**	1,156.7	79.6	11/24/2004	1200	**62.80**	**21.43**	719.5	190.1
10/16/2004	1200	**25.39**	**11.22**	1,087.4	97.9	11/25/2004	1200	**50.73**	**39.18**	757.8	353.0
10/17/2004	1200	**23.48**	**13.98**	1,027.3	122.7	11/26/2004	1200	**36.50**	**110.67**	813.7	1,024.0
10/18/2004	1200	**20.98**	**15.30**	978.0	134.6	11/27/2004	1200	**34.68**	**103.19**	896.7	953.1
10/19/2004	1200	**17.66**	**14.92**	931.7	131.1	11/28/2004	1200	**47.22**	**78.40**	1,033.3	719.0
10/20/2004	1200	**18.84**	**14.08**	893.3	123.6	11/29/2004	1200	**36.50**	**96.21**	1,155.1	887.0
10/21/2004	1200	**23.48**	**13.60**	851.9	119.2	11/30/2004	1200	**32.84**	**95.71**	1,210.0	882.2
10/22/2004	1200	**25.39**	**13.41**	821.5	117.6	12/1/2004	1200	**34.68**	**83.25**	1,370.0	764.7
10/23/2004	1200	**23.48**	**13.31**	796.5	116.6	12/2/2004	1200	**32.84**	**104.19**	1,408.4	962.5
10/24/2004	1200	**23.48**	**12.94**	764.6	113.3	12/3/2004	1200	**32.84**	**104.92**	1,485.5	969.4
10/25/2004	1200	**25.39**	**12.53**	738.8	109.6	12/4/2004	1200	**31.00**	**88.28**	1,573.4	812.1
10/26/2004	1200	**25.39**	**12.00**	718.9	104.9	12/5/2004	1200	**31.00**	**72.53**	1,662.5	663.8
10/27/2004	1200	**29.14**	**11.56**	709.6	100.9	12/6/2004	1200	**31.00**	**56.50**	1,734.0	513.8
10/28/2004	1200	**27.27**	**11.09**	707.4	96.7	12/7/2004	1200	**31.00**	**50.40**	1,761.2	457.0
10/29/2004	1200	**27.27**	**10.73**	715.2	93.5	12/8/2004	1200	**31.00**	**51.92**	1,732.6	471.1
10/30/2004	1200	**27.27**	**10.55**	706.7	91.9	12/9/2004	1200	**31.00**	**59.19**	1,648.6	538.9
10/31/2004	1200	**29.14**	**11.06**	696.2	96.5	12/10/2004	1200	**79.59**	**58.80**	1,551.9	535.3
11/1/2004	1200	**29.14**	**13.05**	683.1	114.3	12/11/2004	1200	**45.45**	**72.42**	1,510.6	662.9
11/2/2004	1200	**27.27**	**15.71**	673.7	138.3	12/12/2004	1200	**34.68**	**92.95**	1,469.4	856.2
11/3/2004	1200	**27.27**	**20.59**	666.5	182.5	12/13/2004	1200	**32.84**	**81.42**	1,408.5	747.4
11/4/2004	1200	**31.00**	**30.02**	660.2	268.6	12/14/2004	1200	**29.14**	**66.78**	1,330.6	610.0
11/5/2004	1200	**34.68**	**32.83**	653.7	294.4	12/15/2004	1200	**29.14**	**54.01**	1,280.6	490.6
11/6/2004	1200	**29.14**	**35.76**	650.4	321.4	12/16/2004	1200	**29.14**	**46.57**	1,278.7	421.4
11/7/2004	1200	**27.27**	**36.50**	652.9	328.3	12/17/2004	1200	**29.14**	**42.82**	1,300.8	386.7
11/8/2004	1200	**27.27**	**40.73**	656.2	367.3	12/18/2004	1200	**29.14**	**36.43**	1,308.3	327.6
11/9/2004	1200	**25.39**	**41.19**	667.9	371.6	12/19/2004	1200	**29.14**	**34.40**	1,275.8	308.9

Appendix 1-C.   Daily mean streamflow for selected sites from October 1, 2004, to October 1, 2009, used as input into the load estimation model at each site.—Continued

[ft³/s, cubic feet per second; Bolded values were estimated using MOVE.1 regression technique]

Date	Time	Daily mean streamflow (ft³/s)				Date	Time	Daily mean streamflow (ft³/s)			
		McTier Creek 02172305	Fishing Brook 0131199050	Edisto River 02175000	Hudson River 01312000			McTier Creek 02172305	Fishing Brook 0131199050	Edisto River 02175000	Hudson River 01312000
12/20/2004	1200	**31.00**	**31.49**	1,209.7	282.1	1/29/2005	1200	**31.00**	**21.04**	1,361.9	186.6
12/21/2004	1200	**29.14**	**28.67**	1,147.7	256.3	1/30/2005	1200	**62.80**	**20.16**	1,280.7	178.6
12/22/2004	1200	**29.14**	**27.68**	1,110.1	247.2	1/31/2005	1200	**45.45**	**19.36**	1,236.6	171.3
12/23/2004	1200	**34.68**	**28.54**	1,094.9	255.0	2/1/2005	1200	**38.31**	**18.69**	1,211.7	165.2
12/24/2004	1200	**34.68**	**73.11**	1,076.1	669.3	2/2/2005	1200	**34.68**	**18.06**	1,211.6	159.6
12/25/2004	1200	**31.00**	**114.49**	1,054.0	1,060.2	2/3/2005	1200	**71.25**	**17.50**	1,249.7	154.5
12/26/2004	1200	**34.68**	**100.33**	1,093.6	925.9	2/4/2005	1200	**54.21**	**17.09**	1,354.3	150.7
12/27/2004	1200	**34.68**	**82.97**	1,201.8	762.0	2/5/2005	1200	**40.11**	**16.64**	1,445.2	146.7
12/28/2004	1200	**31.00**	**66.54**	1,267.6	607.7	2/6/2005	1200	**36.50**	**16.28**	1,496.4	143.4
12/29/2004	1200	**29.14**	**57.05**	1,304.1	519.0	2/7/2005	1200	**34.68**	**16.03**	1,524.0	141.1
12/30/2004	1200	**31.00**	**49.73**	1,330.1	450.8	2/8/2005	1200	**34.68**	**16.16**	1,551.1	142.3
12/31/2004	1200	**31.00**	**44.31**	1,358.5	400.5	2/9/2005	1200	**45.45**	**16.92**	1,593.9	149.2
1/1/2005	1200	**31.00**	**44.77**	1,394.9	404.7	2/10/2005	1200	**50.73**	**18.72**	1,653.9	165.5
1/2/2005	1200	**29.14**	**51.24**	1,426.0	464.9	2/11/2005	1200	**38.31**	**19.28**	1,706.0	170.6
1/3/2005	1200	**29.14**	**61.32**	1,450.6	558.9	2/12/2005	1200	**34.68**	**19.27**	1,734.8	170.5
1/4/2005	1200	**29.14**	**71.32**	1,469.2	652.5	2/13/2005	1200	**34.68**	**19.13**	1,742.7	169.2
1/5/2005	1200	**29.14**	**71.84**	1,469.3	657.4	2/14/2005	1200	**34.68**	**18.51**	1,736.7	163.6
1/6/2005	1200	**27.27**	**64.15**	1,443.7	585.3	2/15/2005	1200	**38.31**	**19.00**	1,750.7	168.0
1/7/2005	1200	**29.14**	**56.23**	1,397.2	511.3	2/16/2005	1200	**34.68**	**19.32**	1,772.9	170.9
1/8/2005	1200	**29.14**	**49.09**	1,340.3	444.8	2/17/2005	1200	**32.84**	**19.80**	1,773.7	175.3
1/9/2005	1200	**29.14**	**43.50**	1,276.8	392.9	2/18/2005	1200	**31.00**	**19.85**	1,735.4	175.7
1/10/2005	1200	**29.14**	**39.29**	1,212.2	354.0	2/19/2005	1200	**29.14**	**19.21**	1,665.4	169.9
1/11/2005	1200	**29.14**	**35.72**	1,155.1	321.1	2/20/2005	1200	**31.00**	**18.65**	1,597.4	164.9
1/12/2005	1200	**31.00**	**33.15**	1,112.4	297.4	2/21/2005	1200	**40.11**	**18.36**	1,568.8	162.2
1/13/2005	1200	**31.00**	**31.83**	1,081.3	285.2	2/22/2005	1200	**54.21**	**18.15**	1,660.3	160.3
1/14/2005	1200	**91.09**	**43.15**	1,126.1	389.7	2/23/2005	1200	**43.68**	**17.83**	1,752.4	157.4
1/15/2005	1200	**48.98**	**101.01**	1,236.0	932.4	2/24/2005	1200	**110.44**	**17.20**	1,757.4	151.7
1/16/2005	1200	**38.31**	**97.27**	1,314.4	897.0	2/25/2005	1200	**62.80**	**16.67**	1,793.8	147.0
1/17/2005	1200	**34.68**	**75.76**	1,370.7	694.2	2/26/2005	1200	**41.90**	**16.09**	1,850.3	141.6
1/18/2005	1200	**34.68**	**57.41**	1,424.4	522.3	2/27/2005	1200	**41.90**	**15.52**	1,905.1	136.6
1/19/2005	1200	**32.84**	**45.10**	1,480.9	407.8	2/28/2005	1200	**77.93**	**15.04**	2,120.4	132.2
1/20/2005	1200	**31.00**	**38.92**	1,528.3	350.6	3/1/2005	1200	**48.98**	**15.15**	2,417.2	133.2
1/21/2005	1200	**31.00**	**34.73**	1,548.6	311.9	3/2/2005	1200	**40.11**	**15.22**	2,613.5	133.8
1/22/2005	1200	**31.00**	**31.31**	1,546.5	280.5	3/3/2005	1200	**36.50**	**14.87**	2,699.1	130.7
1/23/2005	1200	**31.00**	**29.04**	1,530.7	259.7	3/4/2005	1200	**36.50**	**14.57**	2,728.4	128.0
1/24/2005	1200	**29.14**	**26.97**	1,519.4	240.7	3/5/2005	1200	**34.68**	**14.21**	2,763.3	124.8
1/25/2005	1200	**29.14**	**25.60**	1,515.4	228.2	3/6/2005	1200	**34.68**	**13.95**	2,826.2	122.4
1/26/2005	1200	**29.14**	**24.56**	1,528.2	218.6	3/7/2005	1200	**32.84**	**13.84**	2,879.8	121.4
1/27/2005	1200	**27.27**	**23.41**	1,524.8	208.2	3/8/2005	1200	**71.25**	**14.72**	2,905.8	129.3
1/28/2005	1200	**27.27**	**22.38**	1,466.1	198.7	3/9/2005	1200	**45.45**	**15.78**	2,864.3	138.9

Appendix 1-C.   Daily mean streamflow for selected sites from October 1, 2004, to October 1, 2009, used as input into the load estimation model at each site.—Continued

[ft³/s, cubic feet per second; Bolded values were estimated using MOVE.1 regression technique]

		Daily mean streamflow (ft³/s)						Daily mean streamflow (ft³/s)			
Date	Time	McTier Creek 02172305	Fishing Brook 0131199050	Edisto River 02175000	Hudson River 01312000	Date	Time	McTier Creek 02172305	Fishing Brook 0131199050	Edisto River 02175000	Hudson River 01312000
3/10/2005	1200	**38.31**	**16.38**	2,767.0	144.3	4/19/2005	1200	**32.84**	**75.42**	2,077.8	691.0
3/11/2005	1200	**34.68**	**16.45**	2,598.3	145.0	4/20/2005	1200	**31.00**	**85.45**	2,033.2	785.4
3/12/2005	1200	**32.84**	**16.56**	2,420.3	145.9	4/21/2005	1200	**29.14**	**142.91**	2,010.0	1,331.0
3/13/2005	1200	**32.84**	**16.22**	2,255.8	142.9	4/22/2005	1200	**32.84**	**160.19**	1,940.0	1,496.3
3/14/2005	1200	**32.84**	**15.74**	2,108.7	138.5	4/23/2005	1200	**45.45**	**142.18**	1,868.2	1,324.0
3/15/2005	1200	**32.84**	**15.28**	1,971.1	134.4	4/24/2005	1200	**34.68**	**220.60**	1,743.2	2,077.5
3/16/2005	1200	**45.45**	**14.70**	1,891.8	129.1	4/25/2005	1200	**31.00**	**358.34**	1,639.9	3,416.9
3/17/2005	1200	**48.98**	**14.16**	1,903.5	124.3	4/26/2005	1200	**31.00**	**275.14**	1,559.4	2,605.8
3/18/2005	1200	**40.11**	**13.87**	1,950.1	121.6	4/27/2005	1200	**31.00**	**190.94**	1,489.5	1,791.6
3/19/2005	1200	**36.50**	**13.48**	1,989.1	118.2	4/28/2005	1200	**29.14**	**219.23**	1,413.5	2,064.3
3/20/2005	1200	**34.68**	**13.18**	2,033.8	115.5	4/29/2005	1200	**29.14**	**219.88**	1,350.0	2,070.5
3/21/2005	1200	**32.84**	**13.17**	2,086.0	115.4	4/30/2005	1200	**38.31**	**165.18**	1,310.0	1,544.1
3/22/2005	1200	**32.84**	**13.20**	2,159.0	115.7	5/1/2005	1200	**40.11**	**142.45**	1,280.0	1,326.5
3/23/2005	1200	**45.45**	**13.40**	2,373.4	117.5	5/2/2005	1200	**32.84**	**131.63**	1,273.1	1,223.3
3/24/2005	1200	**40.11**	**13.73**	2,606.5	120.4	5/3/2005	1200	**29.14**	**108.87**	1,263.3	1,006.9
3/25/2005	1200	**34.68**	**14.18**	2,735.6	124.5	5/4/2005	1200	**27.27**	**86.47**	1,261.8	795.0
3/26/2005	1200	**32.84**	**14.26**	2,773.5	125.2	5/5/2005	1200	**29.14**	**70.45**	1,264.2	644.3
3/27/2005	1200	**178.51**	**14.68**	2,974.8	128.9	5/6/2005	1200	**29.14**	**58.98**	1,266.8	537.0
3/28/2005	1200	**205.42**	**16.53**	3,568.4	145.7	5/7/2005	1200	**27.27**	**51.43**	1,229.1	466.6
3/29/2005	1200	**76.27**	**21.80**	4,316.4	193.5	5/8/2005	1200	**25.39**	**47.06**	1,172.6	425.9
3/30/2005	1200	**50.73**	**33.72**	4,722.8	302.7	5/9/2005	1200	**23.48**	**43.82**	1,120.0	395.9
3/31/2005	1200	**50.73**	**55.95**	4,747.8	508.7	5/10/2005	1200	**25.39**	**43.70**	1,065.7	394.8
4/1/2005	1200	**61.09**	**108.14**	4,853.3	1,000.0	5/11/2005	1200	**36.50**	**48.63**	1,018.6	440.6
4/2/2005	1200	**62.80**	**179.40**	5,137.2	1,680.6	5/12/2005	1200	**29.14**	**52.45**	976.1	476.1
4/3/2005	1200	**47.22**	**292.64**	5,374.0	2,776.0	5/13/2005	1200	**25.39**	**45.27**	931.4	409.4
4/4/2005	1200	**41.90**	**421.05**	5,454.0	4,031.6	5/14/2005	1200	**23.48**	**36.40**	945.0	327.3
4/5/2005	1200	**38.31**	**318.54**	5,336.9	3,028.2	5/15/2005	1200	**23.48**	**32.73**	998.7	293.5
4/6/2005	1200	**38.31**	**218.11**	5,095.9	2,053.5	5/16/2005	1200	**21.56**	**35.92**	1,029.6	322.9
4/7/2005	1200	**54.21**	**194.58**	4,891.8	1,826.6	5/17/2005	1200	**21.56**	**35.59**	1,057.3	319.9
4/8/2005	1200	**89.46**	**217.98**	4,764.5	2,052.2	5/18/2005	1200	**23.48**	**31.87**	1,113.2	285.6
4/9/2005	1200	**74.60**	**222.27**	4,592.5	2,093.7	5/19/2005	1200	**21.56**	**28.46**	1,177.2	254.4
4/10/2005	1200	**50.73**	**186.96**	4,362.5	1,753.2	5/20/2005	1200	**29.14**	**26.01**	1,147.1	231.8
4/11/2005	1200	**41.90**	**162.62**	4,023.4	1,519.5	5/21/2005	1200	**36.50**	**24.34**	1,079.7	216.6
4/12/2005	1200	**40.11**	**138.61**	3,629.4	1,289.9	5/22/2005	1200	**29.14**	**23.17**	1,086.8	205.9
4/13/2005	1200	**55.94**	**115.04**	3,289.5	1,065.5	5/23/2005	1200	**25.39**	**22.93**	1,125.0	203.8
4/14/2005	1200	**47.22**	**94.36**	3,030.8	869.5	5/24/2005	1200	**23.48**	**25.19**	1,157.3	224.4
4/15/2005	1200	**38.31**	**79.40**	2,801.2	728.4	5/25/2005	1200	**21.56**	**29.23**	1,167.7	261.4
4/16/2005	1200	**36.50**	**69.26**	2,573.2	633.2	5/26/2005	1200	**20.79**	**27.69**	1,145.2	247.3
4/17/2005	1200	**34.68**	**63.75**	2,371.0	581.6	5/27/2005	1200	**19.43**	**25.29**	1,100.2	225.3
4/18/2005	1200	**32.84**	**68.12**	2,194.1	622.5	5/28/2005	1200	**19.43**	**23.83**	1,033.7	212.0

Appendix 1-C. Daily mean streamflow for selected sites from October 1, 2004, to October 1, 2009, used as input into the load estimation model at each site.—Continued

[ft³/s, cubic feet per second; Bolded values were estimated using MOVE.1 regression technique]

Date	Time	Daily mean streamflow (ft³/s)				Date	Time	Daily mean streamflow (ft³/s)			
		McTier Creek 02172305	Fishing Brook 0131199050	Edisto River 02175000	Hudson River 01312000			McTier Creek 02172305	Fishing Brook 0131199050	Edisto River 02175000	Hudson River 01312000
5/29/2005	1200	**20.01**	**22.93**	953.9	203.7	7/8/2005	1200	**27.27**	**16.38**	2,855.6	144.3
5/30/2005	1200	**32.84**	**22.46**	912.6	199.5	7/9/2005	1200	**25.39**	**17.10**	2,915.6	150.8
5/31/2005	1200	**36.50**	**21.80**	873.1	193.5	7/10/2005	1200	**27.27**	**27.61**	3,125.3	246.5
6/1/2005	1200	**34.68**	**21.22**	939.6	188.2	7/11/2005	1200	**31.00**	**33.88**	3,552.1	304.1
6/2/2005	1200	**250.85**	**20.22**	1,172.0	179.1	7/12/2005	1200	**29.14**	**31.26**	3,891.2	280.0
6/3/2005	1200	**119.96**	**19.27**	1,636.8	170.5	7/13/2005	1200	**92.72**	**26.78**	3,827.4	238.9
6/4/2005	1200	**64.50**	**18.22**	2,243.7	161.0	7/14/2005	1200	**108.85**	**25.62**	3,638.7	228.3
6/5/2005	1200	**41.90**	**17.06**	2,647.3	150.5	7/15/2005	1200	**48.98**	**27.40**	3,479.5	244.6
6/6/2005	1200	**32.84**	**15.90**	3,109.4	140.0	7/16/2005	1200	**36.50**	**28.39**	3,197.6	253.7
6/7/2005	1200	**29.14**	**15.00**	3,879.6	131.8	7/17/2005	1200	**31.00**	**29.48**	3,132.3	263.7
6/8/2005	1200	**31.00**	**14.23**	4,373.9	125.0	7/18/2005	1200	**31.00**	**29.81**	3,296.7	266.7
6/9/2005	1200	**29.14**	**14.28**	4,395.8	125.4	7/19/2005	1200	**34.68**	**28.31**	3,049.4	252.9
6/10/2005	1200	**29.14**	**18.52**	4,374.4	163.7	7/20/2005	1200	**29.14**	**26.80**	2,687.5	239.1
6/11/2005	1200	**32.84**	**32.55**	4,432.2	291.9	7/21/2005	1200	**25.39**	**24.08**	2,438.5	214.3
6/12/2005	1200	**31.00**	**40.51**	4,292.5	365.3	7/22/2005	1200	**23.48**	**21.17**	2,277.3	187.8
6/13/2005	1200	**36.50**	**33.00**	3,984.6	296.0	7/23/2005	1200	**21.56**	**18.83**	2,287.5	166.5
6/14/2005	1200	**31.00**	**33.61**	3,714.9	301.6	7/24/2005	1200	**21.37**	**16.74**	2,424.0	147.6
6/15/2005	1200	**25.39**	**56.06**	3,465.3	509.7	7/25/2005	1200	**20.98**	**15.03**	2,481.4	132.1
6/16/2005	1200	**20.59**	**68.38**	3,192.6	624.9	7/26/2005	1200	**21.56**	**13.74**	2,499.4	120.5
6/17/2005	1200	**19.62**	**153.72**	2,896.8	1,434.3	7/27/2005	1200	**19.62**	**19.46**	2,443.9	172.2
6/18/2005	1200	**18.64**	**235.32**	2,581.1	2,219.8	7/28/2005	1200	**19.82**	**56.12**	2,266.4	510.3
6/19/2005	1200	**21.56**	**189.83**	2,317.6	1,780.9	7/29/2005	1200	**36.50**	**55.66**	2,015.7	506.0
6/20/2005	1200	**20.21**	**130.22**	2,104.0	1,209.9	7/30/2005	1200	**32.84**	**40.40**	1,810.4	364.3
6/21/2005	1200	**20.21**	**87.90**	1,887.0	808.5	7/31/2005	1200	**29.14**	**31.26**	1,700.5	280.0
6/22/2005	1200	**21.37**	**62.89**	1,682.5	573.5	8/1/2005	1200	**29.14**	**25.80**	1,945.1	229.9
6/23/2005	1200	**19.62**	**47.60**	1,466.3	431.0	8/2/2005	1200	**25.39**	**23.56**	2,190.2	209.5
6/24/2005	1200	**18.25**	**37.74**	1,269.1	339.7	8/3/2005	1200	**21.56**	**21.51**	2,238.2	190.8
6/25/2005	1200	**18.05**	**31.15**	1,137.2	279.0	8/4/2005	1200	**21.18**	**19.05**	2,245.3	168.5
6/26/2005	1200	**36.50**	**26.42**	1,044.4	235.6	8/5/2005	1200	**19.82**	**16.91**	2,228.6	149.1
6/27/2005	1200	**184.53**	**22.66**	1,086.6	201.3	8/6/2005	1200	**20.59**	**15.09**	2,100.1	132.7
6/28/2005	1200	**197.99**	**19.87**	1,133.2	175.9	8/7/2005	1200	**23.48**	**13.49**	1,800.9	118.3
6/29/2005	1200	**77.93**	**19.12**	1,616.0	169.1	8/8/2005	1200	**32.84**	**12.25**	1,547.5	107.1
6/30/2005	1200	**47.22**	**28.25**	2,225.6	252.4	8/9/2005	1200	**38.31**	**11.26**	1,483.4	98.3
7/1/2005	1200	**36.50**	**31.34**	2,547.3	280.7	8/10/2005	1200	**36.50**	**10.38**	1,587.5	90.4
7/2/2005	1200	**31.00**	**27.98**	2,544.5	249.9	8/11/2005	1200	**40.11**	**9.60**	1,659.2	83.5
7/3/2005	1200	**64.50**	**23.81**	2,524.1	211.8	8/12/2005	1200	**29.14**	**9.01**	1,654.0	78.1
7/4/2005	1200	**66.19**	**20.32**	2,878.3	180.0	8/13/2005	1200	**62.80**	**8.68**	1,694.5	75.2
7/5/2005	1200	**40.11**	**17.98**	2,939.3	158.8	8/14/2005	1200	**31.00**	**8.77**	1,759.3	76.1
7/6/2005	1200	**32.84**	**17.11**	2,868.9	151.0	8/15/2005	1200	**25.39**	**9.39**	1,716.9	81.5
7/7/2005	1200	**31.00**	**17.32**	2,822.8	152.8	8/16/2005	1200	**23.48**	**9.36**	1,660.3	81.2

Appendix 1-C. Daily mean streamflow for selected sites from October 1, 2004, to October 1, 2009, used as input into the load estimation model at each site.—Continued

[ft³/s, cubic feet per second; Bolded values were estimated using MOVE.1 regression technique]

Date	Time	Daily mean streamflow (ft³/s)				Date	Time	Daily mean streamflow (ft³/s)			
		McTier Creek 02172305	Fishing Brook 0131199050	Edisto River 02175000	Hudson River 01312000			McTier Creek 02172305	Fishing Brook 0131199050	Edisto River 02175000	Hudson River 01312000
8/17/2005	1200	**25.39**	**8.87**	1,679.8	76.9	9/26/2005	1200	**17.66**	**9.50**	456.1	82.5
8/18/2005	1200	**29.14**	**8.21**	1,725.3	71.1	9/27/2005	1200	**19.23**	**35.62**	450.3	320.1
8/19/2005	1200	**23.48**	**7.64**	1,708.0	66.0	9/28/2005	1200	**18.84**	**48.87**	445.8	442.8
8/20/2005	1200	**21.56**	**7.90**	1,652.6	68.3	9/29/2005	1200	**19.03**	**36.67**	452.8	329.8
8/21/2005	1200	**20.21**	**10.89**	1,551.5	95.0	9/30/2005	1200	**19.23**	**42.62**	447.6	384.8
8/22/2005	1200 *	**66.19**	**12.58**	1,497.0	110.1	10/1/2005	1200	**18.84**	**39.85**	444.1	359.2
8/23/2005	1200	**102.43**	**11.90**	1,485.7	103.9	10/2/2005	1200	**17.66**	**31.51**	442.6	282.3
8/24/2005	1200	**164.88**	**10.79**	1,480.6	94.1	10/3/2005	1200	**16.87**	**25.85**	440.5	230.4
8/25/2005	1200	**61.09**	**9.74**	1,515.4	84.7	10/4/2005	1200	**17.06**	**21.99**	438.4	195.2
8/26/2005	1200	**34.68**	**8.89**	1,546.3	77.1	10/5/2005	1200	**17.26**	**19.25**	444.3	170.3
8/27/2005	1200	**29.14**	**8.13**	1,639.6	70.4	10/6/2005	1200	**43.68**	**17.12**	476.5	151.0
8/28/2005	1200	**27.27**	**7.96**	1,773.5	68.8	10/7/2005	1200	**64.50**	**15.76**	510.1	138.7
8/29/2005	1200	**25.39**	**7.81**	1,912.2	67.5	10/8/2005	1200	**132.52**	**48.36**	579.5	438.1
8/30/2005	1200	**25.39**	**7.73**	2,073.7	66.8	10/9/2005	1200	**45.45**	**125.51**	653.4	1,165.1
8/31/2005	1200	**27.27**	**23.50**	2,188.7	209.0	10/10/2005	1200	**84.54**	**107.97**	730.9	998.3
9/1/2005	1200	**25.39**	**140.28**	2,174.2	1,305.9	10/11/2005	1200	**41.90**	**75.39**	823.9	690.7
9/2/2005	1200	**23.48**	**165.45**	2,012.9	1,546.7	10/12/2005	1200	**32.84**	**56.51**	887.0	513.9
9/3/2005	1200	**21.56**	**112.25**	1,764.1	1,039.0	10/13/2005	1200	**29.14**	**45.57**	921.9	412.2
9/4/2005	1200	**20.40**	**74.01**	1,543.3	677.8	10/14/2005	1200	**27.27**	**42.17**	929.8	380.7
9/5/2005	1200	**19.43**	**53.57**	1,398.6	486.5	10/15/2005	1200	**25.39**	**101.22**	940.2	934.4
9/6/2005	1200	**19.23**	**40.61**	1,255.3	366.2	10/16/2005	1200	**23.48**	**193.90**	962.9	1,820.1
9/7/2005	1200	**19.62**	**32.38**	1,073.0	290.3	10/17/2005	1200	**23.48**	**248.98**	997.6	2,352.1
9/8/2005	1200	**19.43**	**26.89**	925.5	239.9	10/18/2005	1200	**23.48**	**252.23**	1,056.3	2,383.5
9/9/2005	1200	**18.84**	**23.27**	825.5	206.9	10/19/2005	1200	**21.56**	**213.80**	1,118.2	2,011.9
9/10/2005	1200	**18.25**	**20.49**	755.9	181.6	10/20/2005	1200	**21.56**	**189.76**	1,162.1	1,780.2
9/11/2005	1200	**17.46**	**17.85**	703.8	157.6	10/21/2005	1200	**23.48**	**142.64**	1,187.1	1,328.3
9/12/2005	1200	**17.26**	**15.92**	660.1	140.1	10/22/2005	1200	**23.48**	**106.74**	1,170.2	986.7
9/13/2005	1200	**17.06**	**14.48**	622.9	127.2	10/23/2005	1200	**21.56**	**95.23**	1,083.0	877.7
9/14/2005	1200	**17.26**	**13.12**	594.0	114.9	10/24/2005	1200	**21.56**	**107.21**	981.7	991.1
9/15/2005	1200	**17.66**	**12.14**	567.6	106.1	10/25/2005	1200	**21.56**	**109.71**	889.3	1,014.9
9/16/2005	1200	**17.06**	**11.46**	546.0	100.1	10/26/2005	1200	**21.56**	**121.67**	817.5	1,128.5
9/17/2005	1200	**16.87**	**11.47**	529.6	100.1	10/27/2005	1200	**21.56**	**113.94**	759.2	1,055.0
9/18/2005	1200	**16.67**	**11.52**	517.5	100.6	10/28/2005	1200	**21.56**	**95.29**	714.8	878.3
9/19/2005	1200	**16.07**	**11.22**	506.6	97.9	10/29/2005	1200	**21.56**	**77.82**	681.6	713.6
9/20/2005	1200	**16.27**	**10.79**	496.7	94.1	10/30/2005	1200	**23.48**	**65.47**	659.7	597.6
9/21/2005	1200	**16.67**	**10.48**	485.4	91.3	10/31/2005	1200	**23.48**	**59.86**	641.9	545.2
9/22/2005	1200	**16.47**	**9.89**	480.9	86.0	11/1/2005	1200	**23.48**	**62.76**	627.3	572.3
9/23/2005	1200	**16.87**	**9.44**	480.1	82.0	11/2/2005	1200	**21.56**	**70.91**	613.3	648.6
9/24/2005	1200	**16.67**	**8.85**	469.2	76.7	11/3/2005	1200	**21.56**	**68.09**	603.3	622.2
9/25/2005	1200	**18.64**	**8.45**	460.9	73.2	11/4/2005	1200	**21.56**	**58.11**	598.2	528.9

Appendix 1-C.   Daily mean streamflow for selected sites from October 1, 2004, to October 1, 2009, used as input into the load estimation model at each site.—Continued

[ft³/s, cubic feet per second; Bolded values were estimated using MOVE.1 regression technique]

		Daily mean streamflow (ft³/s)						Daily mean streamflow (ft³/s)			
Date	Time	McTier Creek 02172305	Fishing Brook 0131199050	Edisto River 02175000	Hudson River 01312000	Date	Time	McTier Creek 02172305	Fishing Brook 0131199050	Edisto River 02175000	Hudson River 01312000
11/5/2005	1200	**23.48**	**53.19**	592.0	483.0	12/15/2005	1200	**55.94**	**27.63**	1,223.8	246.7
11/6/2005	1200	**23.48**	**52.43**	589.2	475.9	12/16/2005	1200	**64.50**	**28.50**	1,235.8	254.7
11/7/2005	1200	**23.48**	**66.86**	585.4	610.7	12/17/2005	1200	**43.68**	**28.85**	1,255.8	257.9
11/8/2005	1200	**23.48**	**78.91**	582.7	723.8	12/18/2005	1200	**62.80**	**28.48**	1,334.3	254.5
11/9/2005	1200	**23.48**	**73.89**	581.9	676.6	12/19/2005	1200	**47.22**	**27.58**	1,411.4	246.3
11/10/2005	1200	**23.48**	**138.18**	582.3	1,285.8	12/20/2005	1200	**40.11**	**26.56**	1,448.6	236.9
11/11/2005	1200	**23.48**	**182.61**	574.7	1,711.4	12/21/2005	1200	**36.50**	**25.31**	1,469.7	225.5
11/12/2005	1200	**23.48**	**148.91**	570.4	1,388.3	12/22/2005	1200	**34.68**	**24.45**	1,480.0	217.7
11/13/2005	1200	**23.48**	**116.38**	569.6	1,078.2	12/23/2005	1200	**34.68**	**24.03**	1,482.5	213.8
11/14/2005	1200	**23.48**	**93.24**	567.7	858.9	12/24/2005	1200	**32.84**	**23.65**	1,475.3	210.4
11/15/2005	1200	**23.48**	**81.95**	566.5	752.5	12/25/2005	1200	**69.57**	**23.86**	1,472.3	212.2
11/16/2005	1200	**25.39**	**125.34**	565.2	1,163.5	12/26/2005	1200	**52.47**	**32.80**	1,474.4	294.1
11/17/2005	1200	**27.27**	**203.11**	558.2	1,908.8	12/27/2005	1200	**40.11**	**46.69**	1,489.3	422.5
11/18/2005	1200	**27.27**	**203.12**	560.3	1,908.9	12/28/2005	1200	**40.11**	**46.94**	1,514.5	424.8
11/19/2005	1200	**25.39**	**152.82**	569.0	1,425.7	12/29/2005	1200	**64.50**	**42.97**	1,564.0	388.1
11/20/2005	1200	**25.39**	**118.16**	576.9	1,095.1	12/30/2005	1200	**41.90**	**46.57**	1,618.1	421.4
11/21/2005	1200	**89.46**	**93.40**	659.8	860.4	12/31/2005	1200	**36.50**	**48.72**	1,676.9	441.4
11/22/2005	1200	**62.80**	**77.83**	792.4	713.6	1/1/2006	1200	**34.68**	**47.07**	1,743.4	426.1
11/23/2005	1200	**36.50**	**68.31**	883.1	624.3	1/2/2006	1200	**104.04**	**43.04**	1,834.8	388.7
11/24/2005	1200	**31.00**	**57.07**	997.6	519.2	1/3/2006	1200	**104.04**	**39.35**	2,034.2	354.6
11/25/2005	1200	**27.27**	**50.37**	1,075.1	456.8	1/4/2006	1200	**55.94**	**35.68**	2,293.4	320.7
11/26/2005	1200	**27.27**	**45.89**	1,134.3	415.1	1/5/2006	1200	**47.22**	**34.70**	2,469.0	311.6
11/27/2005	1200	**27.27**	**41.49**	1,169.2	374.4	1/6/2006	1200	**41.90**	**33.04**	2,595.4	296.4
11/28/2005	1200	**29.14**	**39.89**	1,183.4	359.6	1/7/2006	1200	**40.11**	**30.05**	2,743.0	268.9
11/29/2005	1200	**36.50**	**44.16**	1,200.6	399.0	1/8/2006	1200	**38.31**	**28.76**	2,915.3	257.1
11/30/2005	1200	**32.84**	**139.09**	1,234.0	1,294.5	1/9/2006	1200	**36.50**	**27.79**	3,044.3	248.2
12/1/2005	1200	**31.00**	**241.37**	1,300.0	2,278.4	1/10/2006	1200	**36.50**	**26.68**	3,084.8	238.0
12/2/2005	1200	**29.14**	**193.44**	1,370.8	1,815.6	1/11/2006	1200	**36.50**	**25.29**	3,042.1	225.3
12/3/2005	1200	**29.14**	**141.82**	1,437.0	1,320.6	1/12/2006	1200	**34.68**	**28.04**	2,939.1	250.5
12/4/2005	1200	**29.14**	**109.31**	1,477.5	1,011.0	1/13/2006	1200	**38.31**	**34.29**	2,799.7	307.9
12/5/2005	1200	**59.38**	**85.42**	1,477.2	785.1	1/14/2006	1200	**52.47**	**38.98**	2,689.9	351.2
12/6/2005	1200	**69.57**	**68.78**	1,442.6	628.7	1/15/2006	1200	**40.11**	**60.17**	2,583.9	548.1
12/7/2005	1200	**41.90**	**56.86**	1,361.1	517.2	1/16/2006	1200	**36.50**	**71.44**	2,488.2	653.6
12/8/2005	1200	**36.50**	**47.18**	1,278.8	427.1	1/17/2006	1200	**34.68**	**58.13**	2,394.6	529.0
12/9/2005	1200	**54.21**	**43.09**	1,248.1	389.1	1/18/2006	1200	**71.25**	**69.97**	2,321.8	639.9
12/10/2005	1200	**41.90**	**41.48**	1,244.8	374.2	1/19/2006	1200	**47.22**	**159.98**	2,275.3	1,494.3
12/11/2005	1200	**36.50**	**38.52**	1,245.2	346.9	1/20/2006	1200	**40.11**	**193.23**	2,246.7	1,813.6
12/12/2005	1200	**34.68**	**36.71**	1,239.4	330.1	1/21/2006	1200	**38.31**	**163.66**	2,196.1	1,529.5
12/13/2005	1200	**32.84**	**32.45**	1,230.0	291.0	1/22/2006	1200	**43.68**	**139.52**	2,139.9	1,298.6
12/14/2005	1200	**31.00**	**29.36**	1,222.5	262.5	1/23/2006	1200	**45.45**	**123.49**	2,107.1	1,145.8

Appendix 1-C. Daily mean streamflow for selected sites from October 1, 2004, to October 1, 2009, used as input into the load estimation model at each site.—Continued

[ft³/s, cubic feet per second; Bolded values were estimated using MOVE.1 regression technique]

Date	Time	Daily mean streamflow (ft³/s)				Date	Time	Daily mean streamflow (ft³/s)			
		McTier Creek 02172305	Fishing Brook 0131199050	Edisto River 02175000	Hudson River 01312000			McTier Creek 02172305	Fishing Brook 0131199050	Edisto River 02175000	Hudson River 01312000
1/24/2006	1200	**47.22**	**104.28**	2,094.2	963.4	3/5/2006	1200	**31.00**	**21.58**	3,025.0	191.4
1/25/2006	1200	**41.90**	**85.85**	2,081.8	789.2	3/6/2006	1200	**32.84**	**20.39**	3,078.2	180.7
1/26/2006	1200	**36.50**	**71.08**	2,064.4	650.2	3/7/2006	1200	**29.14**	**19.73**	3,048.8	174.7
1/27/2006	1200	**34.68**	**56.97**	2,048.1	518.2	3/8/2006	1200	**29.14**	**19.12**	2,938.7	169.1
1/28/2006	1200	**34.68**	**50.87**	2,022.6	461.4	3/9/2006	1200	**29.14**	**18.90**	2,778.8	167.1
1/29/2006	1200	**36.50**	**47.23**	1,999.1	427.5	3/10/2006	1200	**29.14**	**20.10**	2,582.7	178.0
1/30/2006	1200	**38.31**	**44.54**	1,984.5	402.6	3/11/2006	1200	**31.00**	**24.10**	2,415.5	214.4
1/31/2006	1200	**34.68**	**46.51**	2,010.3	420.9	3/12/2006	1200	**31.00**	**30.15**	2,239.3	269.8
2/1/2006	1200	**32.84**	**50.87**	2,072.1	461.4	3/13/2006	1200	**31.00**	**37.02**	2,043.3	333.0
2/2/2006	1200	**32.84**	**49.56**	2,137.5	449.2	3/14/2006	1200	**29.14**	**80.24**	1,870.4	736.3
2/3/2006	1200	**32.84**	**49.33**	2,267.8	447.1	3/15/2006	1200	**27.27**	**148.90**	1,741.7	1,388.3
2/4/2006	1200	**34.68**	**64.74**	2,451.9	590.8	3/16/2006	1200	**27.27**	**143.65**	1,650.0	1,338.1
2/5/2006	1200	**32.84**	**92.63**	2,579.9	853.2	3/17/2006	1200	**27.27**	**112.29**	1,582.2	1,039.4
2/6/2006	1200	**32.84**	**129.96**	2,590.8	1,207.4	3/18/2006	1200	**27.27**	**84.10**	1,525.1	772.7
2/7/2006	1200	**41.90**	**123.35**	2,547.8	1,144.5	3/19/2006	1200	**27.27**	**66.35**	1,466.5	605.9
2/8/2006	1200	**34.68**	**100.22**	2,467.7	925.0	3/20/2006	1200	**27.27**	**54.16**	1,410.8	492.1
2/9/2006	1200	**32.84**	**76.84**	2,360.7	704.3	3/21/2006	1200	**71.25**	**45.62**	1,372.4	412.7
2/10/2006	1200	**32.84**	**63.87**	2,245.3	582.7	3/22/2006	1200	**45.45**	**40.81**	1,389.3	368.0
2/11/2006	1200	**36.50**	**53.16**	2,148.7	482.7	3/23/2006	1200	**34.68**	**36.45**	1,465.7	327.8
2/12/2006	1200	**38.31**	**47.10**	2,079.4	426.4	3/24/2006	1200	**31.00**	**33.73**	1,625.6	302.7
2/13/2006	1200	**32.84**	**42.55**	2,013.0	384.1	3/25/2006	1200	**31.00**	**31.97**	1,829.4	286.6
2/14/2006	1200	**31.00**	**39.07**	1,944.6	352.0	3/26/2006	1200	**29.14**	**31.44**	2,056.9	281.7
2/15/2006	1200	**31.00**	**36.69**	1,877.8	330.0	3/27/2006	1200	**29.14**	**30.58**	2,279.8	273.7
2/16/2006	1200	**29.14**	**34.93**	1,836.3	313.8	3/28/2006	1200	**29.14**	**31.89**	2,475.5	285.8
2/17/2006	1200	**31.00**	**36.12**	1,804.5	324.7	3/29/2006	1200	**29.14**	**35.40**	2,613.1	318.1
2/18/2006	1200	**31.00**	**43.05**	1,766.7	388.8	3/30/2006	1200	**29.14**	**41.02**	2,682.0	370.0
2/19/2006	1200	**31.00**	**45.05**	1,719.9	407.4	3/31/2006	1200	**27.27**	**57.16**	2,625.4	520.0
2/20/2006	1200	**32.84**	**42.47**	1,670.3	383.4	4/1/2006	1200	**27.27**	**86.74**	2,475.9	797.6
2/21/2006	1200	**32.84**	**38.73**	1,626.2	348.9	4/2/2006	1200	**27.27**	**156.58**	2,289.3	1,461.7
2/22/2006	1200	**34.68**	**35.60**	1,585.3	320.0	4/3/2006	1200	**27.27**	**162.13**	2,096.3	1,514.9
2/23/2006	1200	**82.89**	**33.16**	1,551.4	297.5	4/4/2006	1200	**25.39**	**149.58**	1,891.7	1,394.7
2/24/2006	1200	**47.22**	**31.07**	1,519.5	278.3	4/5/2006	1200	**25.39**	**143.39**	1,704.1	1,335.6
2/25/2006	1200	**48.98**	**29.18**	1,501.0	260.9	4/6/2006	1200	**25.39**	**120.00**	1,550.7	1,112.6
2/26/2006	1200	**105.64**	**28.25**	1,690.5	252.4	4/7/2006	1200	**25.39**	**98.54**	1,432.4	909.0
2/27/2006	1200	**52.47**	**26.90**	2,055.1	240.0	4/8/2006	1200	**29.14**	**88.77**	1,352.2	816.7
2/28/2006	1200	**41.90**	**25.80**	2,322.9	230.0	4/9/2006	1200	**31.00**	**80.96**	1,361.8	743.1
3/1/2006	1200	**38.31**	**24.71**	2,473.3	220.0	4/10/2006	1200	**29.14**	**72.47**	1,386.6	663.3
3/2/2006	1200	**36.50**	**23.61**	2,601.6	210.0	4/11/2006	1200	**25.39**	**67.24**	1,355.9	614.2
3/3/2006	1200	**34.68**	**23.13**	2,751.2	205.5	4/12/2006	1200	**25.39**	**65.74**	1,308.0	600.2
3/4/2006	1200	**32.84**	**22.76**	2,909.0	202.2	4/13/2006	1200	**25.39**	**81.40**	1,263.1	747.3

Appendix 1-C.   Daily mean streamflow for selected sites from October 1, 2004, to October 1, 2009, used as input into the load estimation model at each site.—Continued

[ft³/s, cubic feet per second; Bolded values were estimated using MOVE.1 regression technique]

Date	Time	Daily mean streamflow (ft³/s)				Date	Time	Daily mean streamflow (ft³/s)			
		McTier Creek 02172305	Fishing Brook 0131199050	Edisto River 02175000	Hudson River 01312000			McTier Creek 02172305	Fishing Brook 0131199050	Edisto River 02175000	Hudson River 01312000
4/14/2006	1200	**25.39**	**112.34**	1,216.3	1,039.8	5/24/2006	1200	**18.25**	**109.55**	641.1	1,013.4
4/15/2006	1200	**23.48**	**115.71**	1,168.7	1,071.8	5/25/2006	1200	**16.87**	**92.16**	606.8	848.7
4/16/2006	1200	**21.56**	**126.19**	1,116.7	1,171.5	5/26/2006	1200	**16.27**	**77.50**	579.3	710.6
4/17/2006	1200	**20.98**	**114.04**	1,073.9	1,055.9	5/27/2006	1200	**15.27**	**69.39**	560.6	634.4
4/18/2006	1200	**19.82**	**91.93**	1,033.1	846.5	5/28/2006	1200	**14.05**	**63.05**	528.4	575.0
4/19/2006	1200	**31.00**	**79.61**	1,070.0	730.4	5/29/2006	1200	**13.24**	**54.95**	505.6	499.4
4/20/2006	1200	**34.68**	**81.39**	1,040.0	747.1	5/30/2006	1200	**13.24**	**50.58**	487.5	458.7
4/21/2006	1200	**27.27**	**80.20**	1,000.0	735.9	5/31/2006	1200	**12.63**	**62.10**	471.4	566.1
4/22/2006	1200	**25.39**	**75.00**	897.1	687.0	6/1/2006	1200	**12.01**	**85.05**	460.8	781.7
4/23/2006	1200	**25.39**	**100.44**	891.2	927.0	6/2/2006	1200	**21.56**	**112.86**	448.4	1,044.7
4/24/2006	1200	**23.48**	**190.54**	885.8	1,787.7	6/3/2006	1200	**64.50**	**112.14**	460.1	1,037.9
4/25/2006	1200	**20.21**	**203.78**	870.1	1,915.2	6/4/2006	1200	**29.14**	**131.34**	465.8	1,220.5
4/26/2006	1200	**20.01**	**168.27**	856.9	1,573.7	6/5/2006	1200	**20.59**	**131.46**	503.2	1,221.7
4/27/2006	1200	**25.39**	**132.70**	868.8	1,233.5	6/6/2006	1200	**17.86**	**111.87**	553.9	1,035.4
4/28/2006	1200	**25.39**	**105.52**	869.4	975.2	6/7/2006	1200	**16.47**	**88.78**	597.0	816.8
4/29/2006	1200	**21.56**	**82.60**	905.7	758.6	6/8/2006	1200	**15.07**	**72.72**	609.8	665.6
4/30/2006	1200	**21.56**	**66.96**	960.6	611.6	6/9/2006	1200	**13.65**	**62.41**	598.7	569.0
5/1/2006	1200	**23.48**	**57.78**	981.5	525.8	6/10/2006	1200	**13.24**	**57.05**	576.6	519.0
5/2/2006	1200	**21.18**	**55.32**	961.9	502.8	6/11/2006	1200	**12.21**	**54.33**	541.7	493.6
5/3/2006	1200	**17.86**	**56.42**	908.8	513.1	6/12/2006	1200	**11.38**	**52.41**	499.4	475.7
5/4/2006	1200	**16.87**	**56.84**	839.6	517.0	6/13/2006	1200	**15.87**	**48.69**	472.3	441.1
5/5/2006	1200	**17.46**	**58.11**	774.4	528.9	6/14/2006	1200	**195.01**	**44.14**	535.0	398.9
5/6/2006	1200	**19.43**	**57.54**	724.0	523.5	6/15/2006	1200	**55.94**	**42.49**	596.3	383.6
5/7/2006	1200	**23.48**	**52.26**	689.8	474.4	6/16/2006	1200	**27.27**	**37.50**	747.4	337.4
5/8/2006	1200	**29.14**	**45.70**	692.7	413.3	6/17/2006	1200	**20.21**	**33.30**	921.1	298.8
5/9/2006	1200	**25.39**	**40.03**	694.0	360.8	6/18/2006	1200	**17.66**	**30.72**	1,070.2	275.0
5/10/2006	1200	**23.48**	**37.60**	727.4	338.4	6/19/2006	1200	**16.47**	**29.22**	1,182.0	261.3
5/11/2006	1200	**21.56**	**36.77**	801.1	330.7	6/20/2006	1200	**16.07**	**33.88**	1,263.1	304.1
5/12/2006	1200	**19.82**	**40.01**	843.8	360.7	6/21/2006	1200	**15.07**	**34.40**	1,330.8	308.9
5/13/2006	1200	**18.84**	**72.58**	831.3	664.3	6/22/2006	1200	**14.46**	**31.07**	1,385.1	278.3
5/14/2006	1200	**31.00**	**111.63**	808.0	1,033.1	6/23/2006	1200	**14.05**	**28.66**	1,413.2	256.2
5/15/2006	1200	**32.84**	**125.33**	830.1	1,163.3	6/24/2006	1200	**13.65**	**26.20**	1,402.2	233.6
5/16/2006	1200	**25.39**	**122.14**	820.3	1,132.9	6/25/2006	1200	**14.46**	**23.45**	1,414.7	208.5
5/17/2006	1200	**21.56**	**125.16**	804.5	1,161.7	6/26/2006	1200	**20.59**	**25.48**	1,463.9	227.1
5/18/2006	1200	**20.01**	**155.55**	787.2	1,451.9	6/27/2006	1200	**144.95**	**56.01**	1,306.5	509.2
5/19/2006	1200	**20.59**	**183.62**	766.0	1,721.1	6/28/2006	1200	**21.56**	**172.25**	1,030.9	1,611.9
5/20/2006	1200	**23.48**	**184.67**	740.1	1,731.2	6/29/2006	1200	**17.66**	**321.66**	1,042.8	3,058.6
5/21/2006	1200	**27.27**	**173.02**	714.8	1,619.3	6/30/2006	1200	**15.67**	**276.79**	982.8	2,621.9
5/22/2006	1200	**23.48**	**150.17**	688.1	1,400.3	7/1/2006	1200	**14.66**	**184.22**	925.5	1,726.9
5/23/2006	1200	**20.21**	**128.70**	666.1	1,195.4	7/2/2006	1200	**14.26**	**137.80**	866.9	1,282.2

Appendix 1-C.   Daily mean streamflow for selected sites from October 1, 2004, to October 1, 2009, used as input into the load estimation model at each site.—Continued

[ft³/s, cubic feet per second; Bolded values were estimated using MOVE.1 regression technique]

		Daily mean streamflow (ft³/s)						Daily mean streamflow (ft³/s)			
Date	Time	McTier Creek 02172305	Fishing Brook 0131199050	Edisto River 02175000	Hudson River 01312000	Date	Time	McTier Creek 02172305	Fishing Brook 0131199050	Edisto River 02175000	Hudson River 01312000
7/3/2006	1200	**13.85**	**116.42**	800.0	1,078.6	8/12/2006	1200	**18.05**	**20.46**	387.5	181.3
7/4/2006	1200	**12.83**	**89.55**	741.3	824.1	8/13/2006	1200	**20.01**	**18.21**	443.4	160.9
7/5/2006	1200	**11.38**	**66.73**	692.1	609.5	8/14/2006	1200	**16.67**	**16.48**	520.9	145.2
7/6/2006	1200	**13.04**	**51.56**	659.2	467.8	8/15/2006	1200	**14.46**	**15.53**	564.3	136.6
7/7/2006	1200	**15.47**	**41.55**	625.3	374.9	8/16/2006	1200	**31.00**	**14.67**	584.0	128.8
7/8/2006	1200	**15.07**	**34.85**	576.4	313.0	8/17/2006	1200	**32.84**	**13.64**	618.6	119.6
7/9/2006	1200	**13.65**	**29.95**	532.6	268.0	8/18/2006	1200	**17.86**	**12.61**	622.2	110.4
7/10/2006	1200	**13.04**	**26.27**	490.7	234.2	8/19/2006	1200	**14.46**	**12.01**	616.8	105.0
7/11/2006	1200	**12.42**	**23.28**	463.9	206.9	8/20/2006	1200	**13.44**	**13.82**	605.3	121.3
7/12/2006	1200	**11.80**	**24.13**	456.4	214.7	8/21/2006	1200	**13.65**	**14.77**	591.7	129.8
7/13/2006	1200	**11.38**	**61.89**	447.5	564.2	8/22/2006	1200	**11.80**	**13.95**	568.0	122.4
7/14/2006	1200	**12.01**	**85.74**	436.9	788.1	8/23/2006	1200	**13.04**	**13.23**	527.7	115.9
7/15/2006	1200	**11.59**	**67.93**	423.7	620.7	8/24/2006	1200	**20.21**	**12.71**	507.8	111.3
7/16/2006	1200	**11.38**	**53.54**	410.0	486.3	8/25/2006	1200	**18.84**	**12.03**	567.0	105.1
7/17/2006	1200	**10.55**	**43.62**	405.4	394.1	8/26/2006	1200	**15.47**	**11.19**	662.3	97.6
7/18/2006	1200	**9.71**	**35.64**	392.0	320.3	8/27/2006	1200	**13.24**	**11.55**	804.2	100.9
7/19/2006	1200	**9.49**	**29.86**	384.1	267.1	8/28/2006	1200	**12.63**	**14.36**	865.0	126.1
7/20/2006	1200	**9.28**	**25.46**	376.4	226.8	8/29/2006	1200	**12.01**	**16.06**	813.3	141.4
7/21/2006	1200	**9.07**	**22.49**	366.7	199.8	8/30/2006	1200	**15.27**	**15.83**	745.9	139.3
7/22/2006	1200	**8.21**	**23.62**	356.8	210.0	8/31/2006	1200	**19.03**	**14.46**	689.2	127.0
7/23/2006	1200	**9.49**	**37.56**	346.8	338.0	9/1/2006	1200	**17.46**	**12.97**	639.9	113.6
7/24/2006	1200	**14.46**	**38.12**	367.6	343.2	9/2/2006	1200	**14.86**	**11.58**	589.5	101.2
7/25/2006	1200	**29.14**	**33.12**	376.2	297.1	9/3/2006	1200	**13.85**	**10.95**	554.7	95.5
7/26/2006	1200	**19.82**	**32.17**	389.5	288.3	9/4/2006	1200	**14.05**	**11.55**	540.5	100.9
7/27/2006	1200	**14.66**	**30.28**	421.8	271.0	9/5/2006	1200	**18.05**	**12.49**	552.3	109.3
7/28/2006	1200	**12.83**	**37.16**	506.5	334.4	9/6/2006	1200	**31.00**	**12.62**	710.9	110.4
7/29/2006	1200	**14.66**	**87.88**	563.7	808.3	9/7/2006	1200	**19.62**	**12.95**	848.1	113.4
7/30/2006	1200	**16.27**	**117.74**	605.8	1,091.2	9/8/2006	1200	**17.06**	**12.82**	878.9	112.3
7/31/2006	1200	**14.66**	**99.44**	586.6	917.5	9/9/2006	1200	**15.27**	**12.18**	900.9	106.5
8/1/2006	1200	**12.83**	**102.77**	567.3	949.1	9/10/2006	1200	**14.86**	**11.96**	914.3	104.5
8/2/2006	1200	**12.01**	**154.40**	560.1	1,440.9	9/11/2006	1200	**15.27**	**11.17**	920.6	97.4
8/3/2006	1200	**10.97**	**119.24**	534.7	1,105.4	9/12/2006	1200	**14.86**	**10.27**	892.4	89.4
8/4/2006	1200	**10.13**	**85.04**	493.7	781.5	9/13/2006	1200	**59.38**	**10.10**	850.3	87.9
8/5/2006	1200	**12.83**	**62.88**	458.5	573.5	9/14/2006	1200	**84.54**	**11.58**	896.8	101.1
8/6/2006	1200	**17.06**	**47.91**	424.4	433.9	9/15/2006	1200	**29.14**	**14.12**	1,002.5	123.9
8/7/2006	1200	**16.27**	**38.75**	399.5	349.0	9/16/2006	1200	**20.98**	**16.36**	1,099.5	144.2
8/8/2006	1200	**13.24**	**32.89**	387.0	295.0	9/17/2006	1200	**18.84**	**16.16**	1,191.9	142.3
8/9/2006	1200	**13.24**	**28.20**	375.2	252.0	9/18/2006	1200	**17.06**	**14.74**	1,274.5	129.5
8/10/2006	1200	**12.63**	**24.55**	395.0	218.5	9/19/2006	1200	**17.06**	**13.43**	1,319.3	117.7
8/11/2006	1200	**12.21**	**22.44**	371.4	199.3	9/20/2006	1200	**15.67**	**12.73**	1,320.6	111.4

Appendix 1-C.   Daily mean streamflow for selected sites from October 1, 2004, to October 1, 2009, used as input into the load estimation model at each site.—Continued

[ft³/s, cubic feet per second; Bolded values were estimated using MOVE.1 regression technique]

Date	Time	Daily mean streamflow (ft³/s)				Date	Time	Daily mean streamflow (ft³/s)			
		McTier Creek 02172305	Fishing Brook 0131199050	Edisto River 02175000	Hudson River 01312000			McTier Creek 02172305	Fishing Brook 0131199050	Edisto River 02175000	Hudson River 01312000
9/21/2006	1200	**14.66**	**12.05**	1,278.2	105.4	10/31/2006	1200	**19.03**	**184.32**	577.4	1,727.9
9/22/2006	1200	**14.86**	**11.54**	1,233.3	100.8	11/1/2006	1200	**18.84**	**148.03**	603.0	1,379.8
9/23/2006	1200	**14.86**	**11.43**	1,193.1	99.8	11/2/2006	1200	**18.64**	**137.21**	617.1	1,276.6
9/24/2006	1200	**15.27**	**11.91**	1,150.4	104.1	11/3/2006	1200	**18.05**	**116.53**	618.1	1,079.7
9/25/2006	1200	**13.85**	**13.05**	1,105.6	114.3	11/4/2006	1200	**18.05**	**93.15**	620.1	858.0
9/26/2006	1200	**13.44**	**13.49**	1,020.3	118.2	11/5/2006	1200	**18.25**	**74.58**	623.8	683.1
9/27/2006	1200	**13.65**	**13.06**	873.7	114.4	11/6/2006	1200	**18.45**	**63.36**	626.0	577.9
9/28/2006	1200	**13.44**	**12.23**	752.5	106.9	11/7/2006	1200	**20.01**	**53.38**	670.8	484.8
9/29/2006	1200	**12.83**	**13.99**	671.4	122.7	11/8/2006	1200	**23.48**	**47.07**	766.5	426.0
9/30/2006	1200	**13.85**	**21.26**	610.4	188.6	11/9/2006	1200	**21.56**	**48.30**	791.0	437.5
10/1/2006	1200	**12.63**	**27.94**	564.0	249.6	11/10/2006	1200	**20.40**	**55.99**	793.6	509.1
10/2/2006	1200	**12.21**	**69.20**	523.1	632.6	11/11/2006	1200	**20.01**	**54.16**	783.1	492.0
10/3/2006	1200	**12.63**	**67.18**	491.8	613.6	11/12/2006	1200	**21.18**	**88.44**	755.2	813.6
10/4/2006	1200	**12.63**	**47.31**	473.1	428.3	11/13/2006	1200	**20.98**	**147.02**	736.4	1,370.3
10/5/2006	1200	**12.83**	**38.35**	459.0	345.4	11/14/2006	1200	**21.18**	**132.65**	726.2	1,233.1
10/6/2006	1200	**12.42**	**36.50**	438.3	328.3	11/15/2006	1200	**21.56**	**120.27**	721.6	1,115.2
10/7/2006	1200	**11.59**	**31.77**	421.0	284.7	11/16/2006	1200	**118.38**	**110.54**	853.3	1,022.8
10/8/2006	1200	**15.47**	**27.46**	414.0	245.1	11/17/2006	1200	**45.45**	**211.01**	1,035.7	1,984.9
10/9/2006	1200	**20.21**	**23.77**	405.6	211.4	11/18/2006	1200	**29.14**	**290.94**	1,222.4	2,759.5
10/10/2006	1200	**20.40**	**21.16**	399.6	187.7	11/19/2006	1200	**25.39**	**217.45**	1,296.1	2,047.1
10/11/2006	1200	**18.05**	**19.00**	395.8	168.0	11/20/2006	1200	**21.56**	**152.95**	1,298.1	1,427.0
10/12/2006	1200	**15.67**	**17.95**	401.1	158.5	11/21/2006	1200	**27.27**	**115.72**	1,291.5	1,071.9
10/13/2006	1200	**13.44**	**17.83**	406.1	157.4	11/22/2006	1200	**94.35**	**89.34**	1,371.8	822.1
10/14/2006	1200	**14.66**	**17.56**	417.3	155.0	11/23/2006	1200	**67.88**	**71.83**	1,513.5	657.3
10/15/2006	1200	**13.44**	**16.64**	428.4	146.7	11/24/2006	1200	**38.31**	**60.35**	1,657.5	549.8
10/16/2006	1200	**14.86**	**15.92**	437.4	140.2	11/25/2006	1200	**32.84**	**51.97**	1,749.4	471.7
10/17/2006	1200	**17.26**	**15.88**	440.8	139.8	11/26/2006	1200	**27.27**	**45.81**	1,808.3	414.4
10/18/2006	1200	**18.05**	**35.60**	441.3	319.9	11/27/2006	1200	**25.39**	**41.53**	1,859.2	374.7
10/19/2006	1200	**17.46**	**92.15**	425.2	848.6	11/28/2006	1200	**25.39**	**38.75**	1,920.6	349.1
10/20/2006	1200	**16.47**	**125.30**	430.4	1,163.1	11/29/2006	1200	**25.39**	**36.60**	1,984.9	329.1
10/21/2006	1200	**15.47**	**224.81**	428.6	2,118.2	11/30/2006	1200	**25.39**	**34.81**	2,027.0	312.6
10/22/2006	1200	**18.64**	**219.96**	446.4	2,071.4	12/1/2006	1200	**25.39**	**39.91**	2,048.5	359.7
10/23/2006	1200	**19.82**	**172.34**	469.3	1,612.8	12/2/2006	1200	**23.48**	**91.35**	2,033.1	841.0
10/24/2006	1200	**18.45**	**151.57**	478.3	1,413.8	12/3/2006	1200	**23.48**	**107.18**	1,993.5	990.8
10/25/2006	1200	**17.46**	**129.24**	502.8	1,200.6	12/4/2006	1200	**23.48**	**83.71**	1,957.5	769.0
10/26/2006	1200	**17.86**	**109.41**	533.7	1,012.0	12/5/2006	1200	**23.48**	**66.53**	1,939.2	607.6
10/27/2006	1200	**25.39**	**88.74**	556.3	816.4	12/6/2006	1200	**23.48**	**55.93**	1,937.7	508.5
10/28/2006	1200	**55.94**	**95.27**	562.6	878.1	12/7/2006	1200	**23.48**	**49.87**	1,914.1	452.1
10/29/2006	1200	**27.27**	**217.52**	550.6	2,047.8	12/8/2006	1200	**21.56**	**43.12**	1,822.5	389.4
10/30/2006	1200	**20.98**	**245.52**	555.2	2,318.6	12/9/2006	1200	**21.56**	**39.04**	1,674.7	351.7

Appendix 1-C.   Daily mean streamflow for selected sites from October 1, 2004, to October 1, 2009, used as input into the load estimation model at each site.—Continued

[ft³/s, cubic feet per second; Bolded values were estimated using MOVE.1 regression technique]

Date	Time	Daily mean streamflow (ft³/s)				Date	Time	Daily mean streamflow (ft³/s)			
		McTier Creek 02172305	Fishing Brook 0131199050	Edisto River 02175000	Hudson River 01312000			McTier Creek 02172305	Fishing Brook 0131199050	Edisto River 02175000	Hudson River 01312000
12/10/2006	1200	**21.56**	**36.52**	1,540.0	328.4	1/19/2007	1200	**38.31**	**53.85**	2,376.4	489.1
12/11/2006	1200	**23.48**	**34.86**	1,431.1	313.1	1/20/2007	1200	**32.84**	**48.85**	2,524.8	442.6
12/12/2006	1200	**23.48**	**33.13**	1,339.1	297.2	1/21/2007	1200	**31.00**	**42.64**	2,569.7	385.0
12/13/2006	1200	**23.48**	**31.82**	1,262.6	285.1	1/22/2007	1200	**99.21**	**39.59**	2,627.8	356.8
12/14/2006	1200	**25.39**	**32.68**	1,198.4	293.0	1/23/2007	1200	**61.09**	**37.27**	2,732.6	335.3
12/15/2006	1200	**23.48**	**34.54**	1,148.0	310.2	1/24/2007	1200	**41.90**	**35.04**	2,788.1	314.8
12/16/2006	1200	**23.48**	**37.47**	1,107.8	337.2	1/25/2007	1200	**36.50**	36.02	2,753.5	289.3
12/17/2006	1200	**23.48**	**39.87**	1,074.6	359.3	1/26/2007	1200	**34.68**	31.92	2,704.1	265.4
12/18/2006	1200	**23.48**	**40.15**	1,039.7	361.9	1/27/2007	1200	**32.84**	30.27	2,664.5	249.1
12/19/2006	1200	**23.48**	**39.52**	1,010.5	356.1	1/28/2007	1200	**36.50**	30.33	2,691.5	240.2
12/20/2006	1200	**23.48**	**36.98**	984.8	332.7	1/29/2007	1200	**32.84**	29.52	2,790.9	229.8
12/21/2006	1200	**25.39**	**34.81**	967.1	312.7	1/30/2007	1200	**32.84**	27.56	2,873.0	217.9
12/22/2006	1200	**154.18**	**32.67**	982.7	293.0	1/31/2007	1200	**31.00**	26.46	2,874.2	206.6
12/23/2006	1200	**167.92**	**39.89**	1,150.0	359.6	2/1/2007	1200	**57.66**	25.99	2,883.6	197.7
12/24/2006	1200	**59.38**	**83.05**	1,397.2	762.8	2/2/2007	1200	**66.19**	26.05	3,182.3	192.0
12/25/2006	1200	**102.43**	**90.50**	1,708.0	833.0	2/3/2007	1200	**41.90**	26.56	3,681.5	188.1
12/26/2006	1200	**79.59**	**81.23**	2,402.3	745.6	2/4/2007	1200	**36.50**	25.29	4,014.9	182.9
12/27/2006	1200	**50.73**	**79.08**	3,139.7	725.4	2/5/2007	1200	**32.84**	24.31	4,078.0	176.0
12/28/2006	1200	**43.68**	**71.11**	3,497.0	650.5	2/6/2007	1200	**31.00**	22.36	3,994.0	169.8
12/29/2006	1200	**38.31**	**58.47**	3,585.2	532.2	2/7/2007	1200	**31.00**	21.29	3,887.7	162.9
12/30/2006	1200	**38.31**	**51.94**	3,666.0	471.3	2/8/2007	1200	**31.00**	21.13	3,798.0	160.1
12/31/2006	1200	**34.68**	**47.19**	3,672.3	427.2	2/9/2007	1200	**31.00**	21.48	3,678.9	159.2
1/1/2007	1200	**36.50**	**45.71**	3,591.0	413.5	2/10/2007	1200	**29.14**	21.28	3,518.8	157.0
1/2/2007	1200	**40.11**	**47.77**	3,485.4	432.6	2/11/2007	1200	**29.14**	21.21	3,326.4	154.7
1/3/2007	1200	**34.68**	**46.77**	3,384.2	423.3	2/12/2007	1200	**29.14**	21.00	3,119.0	152.4
1/4/2007	1200	**32.84**	**44.00**	3,323.7	397.6	2/13/2007	1200	**32.84**	21.09	2,913.0	147.9
1/5/2007	1200	**36.50**	**41.76**	3,272.1	376.8	2/14/2007	1200	**57.66**	25.93	2,754.6	168.5
1/6/2007	1200	**64.50**	**95.28**	3,186.0	878.1	2/15/2007	1200	**36.50**	26.90	2,642.6	228.9
1/7/2007	1200	**50.73**	**298.50**	3,013.2	2,833.0	2/16/2007	1200	**32.84**	26.73	2,534.5	257.1
1/8/2007	1200	**64.50**	**291.78**	2,811.3	2,767.7	2/17/2007	1200	**31.00**	25.77	2,427.3	268.2
1/9/2007	1200	**47.22**	**250.59**	2,617.8	2,367.7	2/18/2007	1200	**31.00**	25.06	2,355.7	252.4
1/10/2007	1200	**38.31**	**194.27**	2,456.1	1,823.6	2/19/2007	1200	**29.14**	22.39	2,311.9	229.5
1/11/2007	1200	**34.68**	**139.77**	2,344.6	1,301.0	2/20/2007	1200	**29.14**	22.57	2,279.9	212.1
1/12/2007	1200	**32.84**	**115.95**	2,273.2	1,074.1	2/21/2007	1200	**48.98**	22.59	2,251.6	197.4
1/13/2007	1200	**32.84**	**100.95**	2,236.9	931.9	2/22/2007	1200	**55.94**	22.76	2,301.4	181.6
1/14/2007	1200	**32.84**	**96.46**	2,217.9	889.4	2/23/2007	1200	**40.11**	21.91	2,381.4	169.3
1/15/2007	1200	**31.00**	**90.17**	2,200.8	829.9	2/24/2007	1200	**32.84**	21.58	2,416.7	158.3
1/16/2007	1200	**31.00**	**84.44**	2,183.8	775.9	2/25/2007	1200	**34.68**	22.08	2,439.8	152.3
1/17/2007	1200	**31.00**	**67.38**	2,169.9	615.6	2/26/2007	1200	**34.68**	22.49	2,485.6	148.4
1/18/2007	1200	**36.50**	**57.54**	2,226.5	523.5	2/27/2007	1200	**29.14**	21.96	2,537.2	144.3

Appendix 1-C.   Daily mean streamflow for selected sites from October 1, 2004, to October 1, 2009, used as input into the load estimation model at each site.—Continued

[ft³/s, cubic feet per second; Bolded values were estimated using MOVE.1 regression technique]

Date	Time	Daily mean streamflow (ft³/s)				Date	Time	Daily mean streamflow (ft³/s)			
		McTier Creek 02172305	Fishing Brook 0131199050	Edisto River 02175000	Hudson River 01312000			McTier Creek 02172305	Fishing Brook 0131199050	Edisto River 02175000	Hudson River 01312000
2/28/2007	1200	**29.14**	20.82	2,556.3	138.3	4/9/2007	1200	**23.48**	68.08	914.2	705.9
3/1/2007	1200	**62.80**	20.16	2,540.3	133.2	4/10/2007	1200	**23.48**	58.87	875.7	586.1
3/2/2007	1200	**276.78**	23.54	2,553.6	141.5	4/11/2007	1200	**27.27**	52.01	857.3	494.3
3/3/2007	1200	**119.96**	24.53	2,646.1	147.4	4/12/2007	1200	**50.73**	58.46	843.5	468.2
3/4/2007	1200	**67.88**	23.47	2,765.3	145.1	4/13/2007	1200	**32.84**	58.07	818.4	454.5
3/5/2007	1200	**54.21**	21.97	2,831.0	141.7	4/14/2007	1200	**27.27**	51.64	807.1	419.2
3/6/2007	1200	**47.22**	20.41	2,894.7	136.9	4/15/2007	1200	**50.73**	51.57	855.6	400.3
3/7/2007	1200	**43.68**	20.10	2,968.8	134.5	4/16/2007	1200	**38.31**	69.35	1,033.5	482.8
3/8/2007	1200	**40.11**	19.44	3,071.0	133.5	4/17/2007	1200	**29.14**	116.10	1,217.4	864.0
3/9/2007	1200	**38.31**	18.98	3,141.6	132.9	4/18/2007	1200	**25.39**	125.39	1,387.6	1,003.5
3/10/2007	1200	**36.50**	19.46	3,136.4	132.8	4/19/2007	1200	**25.39**	139.75	1,562.3	1,116.3
3/11/2007	1200	**34.68**	21.02	3,068.1	132.4	4/20/2007	1200	**25.39**	189.32	1,792.0	1,466.6
3/12/2007	1200	**34.68**	21.30	3,038.0	129.6	4/21/2007	1200	**25.39**	244.48	2,030.1	1,874.3
3/13/2007	1200	**34.68**	21.95	3,134.7	129.1	4/22/2007	1200	**23.48**	316.06	2,201.1	2,357.0
3/14/2007	1200	**34.68**	29.12	3,271.7	139.7	4/23/2007	1200	**21.56**	406.08	2,292.8	2,898.1
3/15/2007	1200	**34.68**	67.61	3,268.1	182.2	4/24/2007	1200	**21.56**	629.64	2,396.3	4,101.0
3/16/2007	1200	**40.11**	104.11	3,100.7	289.5	4/25/2007	1200	**20.98**	449.53	2,559.4	4,492.9
3/17/2007	1200	**38.31**	100.20	2,872.0	541.6	4/26/2007	1200	**20.21**	257.38	2,656.0	3,063.7
3/18/2007	1200	**32.84**	83.82	2,599.6	619.8	4/27/2007	1200	**19.43**	196.51	2,582.7	2,217.2
3/19/2007	1200	**31.00**	72.05	2,354.8	535.7	4/28/2007	1200	**20.40**	200.93	2,363.9	1,937.2
3/20/2007	1200	**32.84**	59.24	2,170.0	446.3	4/29/2007	1200	**21.56**	201.69	2,047.4	1,902.1
3/21/2007	1200	**31.00**	46.59	2,073.6	376.7	4/30/2007	1200	**21.56**	204.75	1,830.0	1,878.1
3/22/2007	1200	**31.00**	43.91	2,043.9	331.5	5/1/2007	1200	**19.82**	186.47	1,370.0	1,805.3
3/23/2007	1200	**29.14**	68.57	2,022.1	338.4	5/2/2007	1200	**19.03**	150.21	1,267.1	1,572.7
3/24/2007	1200	**29.14**	93.51	1,977.3	427.6	5/3/2007	1200	**18.45**	123.99	1,108.5	1,356.9
3/25/2007	1200	**29.14**	96.87	1,914.3	544.4	5/4/2007	1200	**18.25**	106.04	970.4	1,186.9
3/26/2007	1200	**27.27**	97.49	1,845.5	591.4	5/5/2007	1200	**21.37**	91.16	871.4	1,051.0
3/27/2007	1200	**27.27**	130.77	1,771.9	711.9	5/6/2007	1200	**23.48**	78.85	809.2	945.8
3/28/2007	1200	**27.27**	218.06	1,697.6	1,116.0	5/7/2007	1200	**20.59**	69.27	748.7	845.6
3/29/2007	1200	**27.27**	208.33	1,601.0	1,530.8	5/8/2007	1200	**19.03**	63.59	701.0	799.3
3/30/2007	1200	**27.27**	160.20	1,502.7	1,432.1	5/9/2007	1200	**18.84**	63.98	670.4	876.6
3/31/2007	1200	**27.27**	132.65	1,404.1	1,209.9	5/10/2007	1200	**18.45**	74.43	650.3	1,030.1
4/1/2007	1200	**27.27**	120.18	1,320.1	1,064.2	5/11/2007	1200	**17.86**	128.59	631.6	1,281.1
4/2/2007	1200	**27.27**	123.26	1,243.9	1,012.1	5/12/2007	1200	**48.98**	123.10	622.3	1,336.3
4/3/2007	1200	**27.27**	144.33	1,183.2	1,037.3	5/13/2007	1200	**71.25**	88.72	646.2	1,098.5
4/4/2007	1200	**25.39**	174.05	1,156.6	1,210.1	5/14/2007	1200	**29.14**	65.95	681.9	857.6
4/5/2007	1200	**23.48**	164.89	1,120.8	1,316.9	5/15/2007	1200	**21.56**	57.44	676.8	728.2
4/6/2007	1200	**23.48**	124.51	1,067.2	1,151.8	5/16/2007	1200	**20.01**	89.48	625.3	1,015.1
4/7/2007	1200	**23.48**	98.63	1,013.8	969.4	5/17/2007	1200	**19.03**	134.38	581.6	1,986.5
4/8/2007	1200	**23.48**	81.84	963.1	825.8	5/18/2007	1200	**18.45**	112.61	556.7	1,817.5

Appendix 1-C.   Daily mean streamflow for selected sites from October 1, 2004, to October 1, 2009, used as input into the load estimation model at each site.—Continued

[ft³/s, cubic feet per second; Bolded values were estimated using MOVE.1 regression technique]

| Date | Time | Daily mean streamflow (ft³/s) | | | | Date | Time | Daily mean streamflow (ft³/s) | | | |
		McTier Creek 02172305	Fishing Brook 0131199050	Edisto River 02175000	Hudson River 01312000			McTier Creek 02172305	Fishing Brook 0131199050	Edisto River 02175000	Hudson River 01312000
5/19/2007	1200	**17.46**	85.41	537.1	1,277.3	6/28/2007	1200	9.92	9.54	719.4	105.6
5/20/2007	1200	**16.27**	71.43	553.7	987.3	6/29/2007	1200	9.75	7.15	668.4	93.9
5/21/2007	1200	**16.07**	62.90	577.3	886.9	6/30/2007	1200	9.30	6.47	599.1	83.9
5/22/2007	1200	**15.27**	53.88	570.6	791.1	7/1/2007	1200	9.36	6.41	535.0	77.2
5/23/2007	1200	**14.86**	45.44	532.6	660.3	7/2/2007	1200	49.12	8.07	501.7	71.7
5/24/2007	1200	**14.26**	39.02	485.0	565.8	7/3/2007	1200	52.78	5.94	506.1	66.2
5/25/2007	1200	**13.44**	33.17	445.8	526.7	7/4/2007	1200	20.18	6.29	523.3	63.3
5/26/2007	1200	**13.24**	29.08	414.9	484.0	7/5/2007	1200	15.19	8.26	605.7	71.3
5/27/2007	1200	**12.83**	26.21	393.1	420.7	7/6/2007	1200	12.51	9.24	611.8	91.0
5/28/2007	1200	**12.63**	29.89	373.6	393.5	7/7/2007	1200	11.65	9.89	558.3	113.2
5/29/2007	1200	**12.21**	27.94	358.6	376.1	7/8/2007	1200	13.39	8.61	516.3	142.2
5/30/2007	1200	**11.38**	23.59	346.5	322.0	7/9/2007	1200	14.86	12.38	489.2	202.7
5/31/2007	1200	**10.97**	20.73	337.0	278.5	7/10/2007	1200	14.03	15.37	479.1	269.3
6/1/2007	1200	**11.18**	18.81	330.7	248.7	7/11/2007	1200	22.93	16.14	478.3	247.8
6/2/2007	1200	**13.24**	16.60	336.5	225.2	7/12/2007	1200	29.58	25.62	487.7	274.4
6/3/2007	1200	**105.64**	14.81	358.5	202.7	7/13/2007	1200	18.60	25.47	473.7	314.0
6/4/2007	1200	**47.22**	22.97	363.1	190.8	7/14/2007	1200	14.99	17.55	471.2	261.8
6/5/2007	1200	**25.39**	30.92	383.2	209.2	7/15/2007	1200	15.28	14.50	488.4	217.3
6/6/2007	1200	**18.64**	25.92	410.7	224.9	7/16/2007	1200	16.02	12.22	642.2	186.9
6/7/2007	1200	**15.87**	20.54	443.2	203.8	7/17/2007	1200	13.45	10.08	810.4	160.2
6/8/2007	1200	**13.44**	17.40	480.1	181.6	7/18/2007	1200	11.76	9.15	983.7	142.2
6/9/2007	1200	**12.01**	15.39	508.0	170.2	7/19/2007	1200	12.20	12.57	989.7	142.6
6/10/2007	1200	**10.76**	12.40	517.7	152.1	7/20/2007	1200	11.06	30.78	835.1	273.8
6/11/2007	1200	**11.38**	11.27	526.4	136.6	7/21/2007	1200	10.08	55.98	685.8	402.9
6/12/2007	1200	**16.67**	10.00	549.8	123.5	7/22/2007	1200	9.51	52.40	597.4	374.1
6/13/2007	1200	18.63	8.97	615.1	110.1	7/23/2007	1200	8.90	30.30	523.9	296.2
6/14/2007	1200	21.10	8.01	702.6	99.1	7/24/2007	1200	8.32	19.39	466.7	238.6
6/15/2007	1200	26.55	7.26	796.4	90.0	7/25/2007	1200	8.19	13.78	436.1	197.6
6/16/2007	1200	29.49	6.84	857.4	82.8	7/26/2007	1200	8.19	10.79	415.4	166.0
6/17/2007	1200	20.46	7.11	838.3	77.8	7/27/2007	1200	8.45	8.52	401.7	145.2
6/18/2007	1200	16.80	7.70	861.4	73.9	7/28/2007	1200	9.59	8.05	437.4	135.1
6/19/2007	1200	14.38	7.21	940.0	72.5	7/29/2007	1200	11.97	8.54	526.1	126.7
6/20/2007	1200	13.24	27.75	1,002.3	109.7	7/30/2007	1200	13.07	7.97	537.4	112.1
6/21/2007	1200	13.70	43.11	1,054.0	147.5	7/31/2007	1200	14.88	6.83	524.5	98.4
6/22/2007	1200	13.10	40.13	1,048.0	153.5	8/1/2007	1200	11.79	6.10	651.3	86.7
6/23/2007	1200	12.23	34.42	986.7	157.3	8/2/2007	1200	10.07	5.41	723.9	77.4
6/24/2007	1200	11.43	26.54	950.0	158.4	8/3/2007	1200	8.97	4.86	673.6	69.3
6/25/2007	1200	10.87	20.04	930.6	144.2	8/4/2007	1200	8.12	5.35	627.7	63.3
6/26/2007	1200	12.79	15.03	870.3	128.9	8/5/2007	1200	7.32	5.38	657.3	56.4
6/27/2007	1200	10.80	11.65	793.2	115.8	8/6/2007	1200	6.56	5.42	642.1	54.0

Appendix 1-C.   Daily mean streamflow for selected sites from October 1, 2004, to October 1, 2009, used as input into the load estimation model at each site.—Continued

[ft³/s, cubic feet per second; Bolded values were estimated using MOVE.1 regression technique]

		Daily mean streamflow (ft³/s)						Daily mean streamflow (ft³/s)			
Date	Time	McTier Creek 02172305	Fishing Brook 0131199050	Edisto River 02175000	Hudson River 01312000	Date	Time	McTier Creek 02172305	Fishing Brook 0131199050	Edisto River 02175000	Hudson River 01312000
8/7/2007	1200	6.31	5.77	574.3	56.8	9/16/2007	1200	16.07	10.76	544.7	40.6
8/8/2007	1200	6.21	9.12	495.3	67.9	9/17/2007	1200	12.08	10.05	602.6	41.5
8/9/2007	1200	5.56	9.73	434.9	83.6	9/18/2007	1200	10.48	8.80	594.3	40.0
8/10/2007	1200	5.29	8.83	392.3	83.1	9/19/2007	1200	9.75	7.26	587.8	38.2
8/11/2007	1200	4.70	7.70	355.6	75.0	9/20/2007	1200	9.75	5.85	570.3	36.5
8/12/2007	1200	4.50	5.86	329.5	67.5	9/21/2007	1200	10.92	4.64	534.7	34.9
8/13/2007	1200	5.04	4.86	313.8	60.4	9/22/2007	1200	12.06	3.84	494.3	33.1
8/14/2007	1200	4.57	4.85	298.8	54.0	9/23/2007	1200	10.69	3.37	456.0	31.5
8/15/2007	1200	4.31	4.15	290.9	48.6	9/24/2007	1200	9.78	2.58	422.2	29.3
8/16/2007	1200	4.52	3.13	283.1	45.4	9/25/2007	1200	8.96	2.67	402.6	27.4
8/17/2007	1200	5.53	3.37	272.1	45.0	9/26/2007	1200	8.52	4.46	398.8	28.8
8/18/2007	1200	5.11	4.21	257.3	45.3	9/27/2007	1200	8.58	4.77	398.0	30.9
8/19/2007	1200	4.73	4.03	244.9	45.3	9/28/2007	1200	8.69	11.23	387.9	48.6
8/20/2007	1200	4.38	4.38	232.2	42.4	9/29/2007	1200	7.98	14.14	366.3	65.7
8/21/2007	1200	4.22	4.12	224.5	39.3	9/30/2007	1200	6.92	12.66	344.8	67.7
8/22/2007	1200	4.13	3.68	232.5	36.3	10/1/2007	1200	7.08	10.04	323.3	60.7
8/23/2007	1200	4.98	3.84	258.2	34.7	10/2/2007	1200	6.92	7.61	304.3	53.9
8/24/2007	1200	5.15	4.62	274.7	35.4	10/3/2007	1200	8.42	6.19	298.8	48.8
8/25/2007	1200	5.02	4.58	257.5	35.8	10/4/2007	1200	11.49	5.48	299.2	45.1
8/26/2007	1200	11.75	4.26	242.6	34.2	10/5/2007	1200	14.57	5.04	293.7	41.8
8/27/2007	1200	18.30	3.95	250.3	32.3	10/6/2007	1200	13.29	4.74	294.2	40.4
8/28/2007	1200	11.64	3.53	290.4	28.9	10/7/2007	1200	12.27	4.98	296.6	41.8
8/29/2007	1200	10.34	3.32	348.7	26.5	10/8/2007	1200	11.76	10.90	310.6	61.5
8/30/2007	1200	26.80	3.14	403.7	25.0	10/9/2007	1200	10.73	23.60	321.2	79.1
8/31/2007	1200	16.76	2.86	428.8	24.1	10/10/2007	1200	9.71	28.66	326.5	115.5
9/1/2007	1200	13.56	2.50	430.3	22.7	10/11/2007	1200	9.13	28.27	326.3	172.3
9/2/2007	1200	11.64	2.25	455.9	21.7	10/12/2007	1200	8.58	30.72	320.3	176.7
9/3/2007	1200	9.67	2.28	461.5	21.0	10/13/2007	1200	8.58	31.93	312.5	178.4
9/4/2007	1200	8.83	2.23	470.2	20.1	10/14/2007	1200	8.50	33.74	302.1	180.6
9/5/2007	1200	8.22	2.08	450.0	19.4	10/15/2007	1200	8.40	31.07	294.7	177.0
9/6/2007	1200	7.61	2.26	419.6	18.7	10/16/2007	1200	8.40	24.78	288.3	161.3
9/7/2007	1200	7.09	2.54	412.9	18.0	10/17/2007	1200	9.01	19.09	284.9	152.7
9/8/2007	1200	6.94	3.46	420.3	17.8	10/18/2007	1200	9.07	15.95	283.5	142.5
9/9/2007	1200	6.50	4.66	425.8	28.4	10/19/2007	1200	10.05	17.15	284.1	138.2
9/10/2007	1200	6.03	6.47	416.3	41.7	10/20/2007	1200	10.59	63.21	278.5	511.7
9/11/2007	1200	5.78	7.75	389.0	38.7	10/21/2007	1200	9.21	84.49	274.2	896.3
9/12/2007	1200	6.94	10.45	358.0	39.6	10/22/2007	1200	8.84	54.43	278.1	605.2
9/13/2007	1200	11.18	10.01	343.4	39.6	10/23/2007	1200	9.45	53.59	279.5	475.7
9/14/2007	1200	11.03	8.63	395.6	36.8	10/24/2007	1200	17.77	159.75	290.2	1,250.2
9/15/2007	1200	23.55	11.21	451.5	38.2	10/25/2007	1200	34.81	126.12	286.9	1,495.9

Appendix 1-C. Daily mean streamflow for selected sites from October 1, 2004, to October 1, 2009, used as input into the load estimation model at each site.—Continued

[ft³/s, cubic feet per second; Bolded values were estimated using MOVE.1 regression technique]

Date	Time	Daily mean streamflow (ft³/s)				Date	Time	Daily mean streamflow (ft³/s)			
		McTier Creek 02172305	Fishing Brook 0131199050	Edisto River 02175000	Hudson River 01312000			McTier Creek 02172305	Fishing Brook 0131199050	Edisto River 02175000	Hudson River 01312000
10/26/2007	1200	20.27	74.05	303.9	1,040.6	12/5/2007	1200	14.44	39.43	384.3	352.6
10/27/2007	1200	16.27	66.94	318.6	839.7	12/6/2007	1200	14.41	34.81	382.2	308.0
10/28/2007	1200	13.55	114.18	335.9	1,308.0	12/7/2007	1200	16.24	31.70	381.3	279.4
10/29/2007	1200	12.32	105.31	365.8	1,350.3	12/8/2007	1200	17.20	29.94	382.0	258.3
10/30/2007	1200	11.58	73.59	407.4	984.6	12/9/2007	1200	16.37	28.12	380.1	235.4
10/31/2007	1200	12.19	52.56	438.0	731.8	12/10/2007	1200	17.49	27.21	379.1	220.8
11/1/2007	1200	12.23	41.63	447.1	543.6	12/11/2007	1200	17.20	26.84	375.7	209.8
11/2/2007	1200	11.89	33.59	439.0	424.7	12/12/2007	1200	16.06	31.25	373.8	214.6
11/3/2007	1200	11.46	29.08	427.6	343.6	12/13/2007	1200	15.45	32.16	374.1	208.4
11/4/2007	1200	11.47	25.26	410.5	290.2	12/14/2007	1200	16.32	32.18	378.6	206.4
11/5/2007	1200	11.33	22.06	389.8	249.1	12/15/2007	1200	19.11	28.35	379.6	198.9
11/6/2007	1200	11.11	28.18	372.8	244.9	12/16/2007	1200	90.03	30.88	402.1	201.9
11/7/2007	1200	10.24	36.62	355.0	283.5	12/17/2007	1200	50.73	31.15	412.6	203.9
11/8/2007	1200	10.43	36.17	344.9	276.7	12/18/2007	1200	26.13	29.90	445.6	201.9
11/9/2007	1200	11.99	31.23	334.2	248.6	12/19/2007	1200	21.07	27.72	510.8	200.2
11/10/2007	1200	12.38	26.73	327.6	223.8	12/20/2007	1200	19.23	26.14	584.2	192.7
11/11/2007	1200	12.61	23.02	324.2	201.7	12/21/2007	1200	33.78	24.83	678.7	183.6
11/12/2007	1200	13.00	20.89	323.3	184.1	12/22/2007	1200	31.61	24.41	782.8	173.8
11/13/2007	1200	13.64	22.20	325.5	176.4	12/23/2007	1200	24.09	28.16	861.2	171.5
11/14/2007	1200	13.61	23.24	326.2	175.4	12/24/2007	1200	22.97	68.29	912.4	291.5
11/15/2007	1200	14.59	39.34	330.6	239.5	12/25/2007	1200	20.42	103.95	953.0	654.8
11/16/2007	1200	15.20	75.29	328.5	609.3	12/26/2007	1200	33.21	95.98	1,041.2	708.9
11/17/2007	1200	14.68	73.94	325.4	676.8	12/27/2007	1200	28.40	75.42	1,108.5	592.5
11/18/2007	1200	14.48	55.48	324.6	529.2	12/28/2007	1200	23.41	59.38	1,150.3	500.8
11/19/2007	1200	14.64	46.37	325.6	413.3	12/29/2007	1200	28.11	61.53	1,183.5	448.8
11/20/2007	1200	14.88	41.61	327.9	362.5	12/30/2007	1200	76.10	70.00	1,217.9	404.9
11/21/2007	1200	15.03	40.88	335.6	334.6	12/31/2007	1200	91.41	64.91	1,235.9	377.9
11/22/2007	1200	15.32	97.82	332.9	513.8	1/1/2008	1200	43.34	55.09	1,242.9	351.9
11/23/2007	1200	18.19	176.05	338.1	1,175.5	1/2/2008	1200	30.21	48.07	1,270.2	327.4
11/24/2007	1200	16.56	124.42	354.2	1,157.8	1/3/2008	1200	26.00	39.15	1,329.8	282.5
11/25/2007	1200	15.79	89.08	363.7	893.7	1/4/2008	1200	24.38	36.22	1,420.9	261.3
11/26/2007	1200	16.10	67.82	363.3	724.9	1/5/2008	1200	23.91	34.82	1,538.6	253.0
11/27/2007	1200	17.78	104.64	371.0	770.0	1/6/2008	1200	23.63	33.42	1,664.2	243.5
11/28/2007	1200	16.61	133.80	356.8	980.0	1/7/2008	1200	22.92	35.38	1,782.4	235.6
11/29/2007	1200	16.12	103.43	357.6	892.9	1/8/2008	1200	22.17	82.02	1,893.5	302.1
11/30/2007	1200	15.64	78.74	366.7	771.9	1/9/2008	1200	22.12	268.23	1,987.2	955.0
12/1/2007	1200	15.68	60.19	375.1	613.6	1/10/2008	1200	22.57	329.96	2,034.9	1,797.9
12/2/2007	1200	14.64	47.74	379.1	485.0	1/11/2008	1200	46.50	202.91	2,027.0	1,892.3
12/3/2007	1200	15.34	46.39	384.6	432.7	1/12/2008	1200	45.25	196.65	1,986.9	1,732.1
12/4/2007	1200	14.98	42.95	383.1	394.0	1/13/2008	1200	29.83	156.15	1,906.5	1,490.3

Appendix 1-C.   Daily mean streamflow for selected sites from October 1, 2004, to October 1, 2009, used as input into the load estimation model at each site.—Continued

[ft³/s, cubic feet per second; Bolded values were estimated using MOVE.1 regression technique]

Date	Time	Daily mean streamflow (ft³/s)				Date	Time	Daily mean streamflow (ft³/s)			
		McTier Creek 02172305	Fishing Brook 0131199050	Edisto River 02175000	Hudson River 01312000			McTier Creek 02172305	Fishing Brook 0131199050	Edisto River 02175000	Hudson River 01312000
1/14/2008	1200	24.97	114.61	1,782.5	1,185.8	2/23/2008	1200	52.75	46.37	1,989.2	363.4
1/15/2008	1200	22.57	92.20	1,628.8	962.1	2/24/2008	1200	36.05	40.97	2,167.3	324.4
1/16/2008	1200	21.59	68.74	1,475.5	760.7	2/25/2008	1200	30.84	37.20	2,302.3	293.5
1/17/2008	1200	47.51	57.20	1,396.1	607.3	2/26/2008	1200	35.23	36.63	2,386.4	257.1
1/18/2008	1200	48.70	53.80	1,378.2	530.1	2/27/2008	1200	39.22	35.61	2,469.6	232.2
1/19/2008	1200	46.97	49.22	1,400.4	459.8	2/28/2008	1200	30.11	32.13	2,556.2	210.0
1/20/2008	1200	60.09	43.33	1,526.8	395.9	2/29/2008	1200	27.37	31.26	2,645.0	190.0
1/21/2008	1200	36.26	37.62	1,673.2	348.9	3/1/2008	1200	26.14	31.18	2,698.4	180.0
1/22/2008	1200	30.62	36.08	1,791.9	311.3	3/2/2008	1200	25.27	29.07	2,718.9	170.0
1/23/2008	1200	37.08	35.49	1,867.9	293.0	3/3/2008	1200	24.63	28.57	2,714.5	165.0
1/24/2008	1200	31.04	33.91	1,938.5	271.6	3/4/2008	1200	31.44	36.79	2,690.1	195.0
1/25/2008	1200	26.77	31.14	2,026.4	249.5	3/5/2008	1200	68.14	62.11	2,625.2	241.4
1/26/2008	1200	25.31	30.25	2,116.9	238.0	3/6/2008	1200	40.20	85.67	2,336.9	343.7
1/27/2008	1200	24.31	28.99	2,199.3	230.3	3/7/2008	1200	52.04	79.56	2,466.7	447.8
1/28/2008	1200	23.38	28.13	2,249.6	215.0	3/8/2008	1200	74.24	78.93	2,441.0	509.3
1/29/2008	1200	22.40	26.83	2,251.9	202.0	3/9/2008	1200	41.98	148.55	2,420.8	836.1
1/30/2008	1200	33.46	31.03	2,215.5	203.1	3/10/2008	1200	33.10	175.29	2,390.5	1,263.7
1/31/2008	1200	27.95	35.75	2,136.6	211.8	3/11/2008	1200	29.80	141.09	2,347.3	1,335.8
2/1/2008	1200	41.72	35.87	2,034.1	227.7	3/12/2008	1200	28.13	105.61	2,301.3	1,138.8
2/2/2008	1200	42.43	38.18	1,913.8	245.9	3/13/2008	1200	26.31	78.20	2,237.9	921.0
2/3/2008	1200	30.20	37.08	1,782.7	248.9	3/14/2008	1200	26.40	64.40	2,177.5	768.4
2/4/2008	1200	26.55	32.96	1,658.6	241.1	3/15/2008	1200	25.41	59.49	2,153.4	645.5
2/5/2008	1200	24.33	39.02	1,564.8	243.9	3/16/2008	1200	27.75	59.51	2,171.3	568.5
2/6/2008	1200	27.82	74.20	1,511.9	288.6	3/17/2008	1200	25.26	54.08	2,240.1	504.1
2/7/2008	1200	39.66	90.80	1,479.4	385.9	3/18/2008	1200	24.25	48.38	2,294.3	446.3
2/8/2008	1200	28.12	76.32	1,444.7	438.0	3/19/2008	1200	25.68	49.09	2,314.3	420.8
2/9/2008	1200	24.39	60.31	1,414.9	410.2	3/20/2008	1200	43.66	73.79	2,349.3	510.1
2/10/2008	1200	22.30	51.05	1,401.4	373.1	3/21/2008	1200	30.87	83.22	2,392.4	658.3
2/11/2008	1200	21.00	41.05	1,401.2	334.2	3/22/2008	1200	26.45	68.64	2,417.9	651.2
2/12/2008	1200	21.25	41.01	1,412.2	302.2	3/23/2008	1200	24.44	55.70	2,379.1	560.4
2/13/2008	1200	23.29	40.39	1,157.7	288.8	3/24/2008	1200	23.38	47.59	2,297.5	481.9
2/14/2008	1200	22.11	36.37	1,491.2	274.1	3/25/2008	1200	21.99	42.33	2,207.2	420.9
2/15/2008	1200	21.19	34.00	1,473.4	259.1	3/26/2008	1200	22.39	41.95	2,123.5	384.7
2/16/2008	1200	20.83	31.55	1,417.3	242.4	3/27/2008	1200	22.82	39.74	2,032.1	353.9
2/17/2008	1200	21.14	30.29	1,374.3	227.0	3/28/2008	1200	21.86	40.03	1,928.5	338.9
2/18/2008	1200	71.97	47.59	1,416.8	242.2	3/29/2008	1200	22.37	36.15	1,816.9	317.1
2/19/2008	1200	56.15	88.94	1,557.9	323.3	3/30/2008	1200	24.33	33.96	1,719.7	295.5
2/20/2008	1200	33.73	88.89	1,712.8	467.5	3/31/2008	1200	24.63	35.91	1,661.4	287.9
2/21/2008	1200	29.09	68.09	1,799.0	466.2	4/1/2008	1200	23.66	55.87	1,650.4	305.5
2/22/2008	1200	72.21	53.31	1,852.0	410.6	4/2/2008	1200	22.62	111.25	1,667.3	632.9

Appendix 1-C. Daily mean streamflow for selected sites from October 1, 2004, to October 1, 2009, used as input into the load estimation model at each site.—Continued

[ft³/s, cubic feet per second; Bolded values were estimated using MOVE.1 regression technique]

Date	Time	Daily mean streamflow (ft³/s)				Date	Time	Daily mean streamflow (ft³/s)			
		McTier Creek 02172305	Fishing Brook 0131199050	Edisto River 02175000	Hudson River 01312000			McTier Creek 02172305	Fishing Brook 0131199050	Edisto River 02175000	Hudson River 01312000
4/3/2008	1200	23.69	166.62	1,741.8	1,119.7	5/13/2008	1200	32.34	42.98	713.0	527.8
4/4/2008	1200	26.08	136.68	1,837.6	1,122.5	5/14/2008	1200	22.16	36.55	824.8	462.7
4/5/2008	1200	61.06	135.95	1,924.3	1,124.2	5/15/2008	1200	19.25	33.16	985.4	439.0
4/6/2008	1200	78.15	152.55	2,096.6	1,235.8	5/16/2008	1200	18.18	29.59	1,121.1	433.5
4/7/2008	1200	42.84	184.24	2,340.6	1,385.5	5/17/2008	1200	16.78	27.15	1,205.9	397.5
4/8/2008	1200	32.33	216.75	2,558.4	1,613.7	5/18/2008	1200	15.18	25.28	1,189.6	341.7
4/9/2008	1200	28.04	261.21	2,647.6	1,900.3	5/19/2008	1200	13.60	24.68	1,110.3	330.1
4/10/2008	1200	25.96	380.53	2,617.6	2,519.5	5/20/2008	1200	13.11	24.59	1,027.5	329.0
4/11/2008	1200	24.26	372.86	2,552.7	3,025.4	5/21/2008	1200	16.00	23.55	995.8	307.4
4/12/2008	1200	22.96	428.63	2,529.7	3,123.2	5/22/2008	1200	14.00	23.95	1,011.9	295.1
4/13/2008	1200	23.98	508.44	2,504.1	3,948.0	5/23/2008	1200	13.00	32.20	1,090.9	289.3
4/14/2008	1200	21.79	271.43	2,412.6	3,293.3	5/24/2008	1200	11.22	39.74	1,151.9	304.6
4/15/2008	1200	22.18	178.02	2,292.3	2,303.9	5/25/2008	1200	10.83	34.11	1,044.7	313.4
4/16/2008	1200	20.62	164.39	2,159.8	1,850.4	5/26/2008	1200	10.13	28.55	844.5	290.9
4/17/2008	1200	19.51	191.25	2,053.8	1,844.2	5/27/2008	1200	9.61	30.90	696.7	311.2
4/18/2008	1200	20.26	270.22	1,962.3	2,301.5	5/28/2008	1200	12.71	32.91	607.2	335.4
4/19/2008	1200	20.10	349.58	1,849.0	3,008.9	5/29/2008	1200	20.13	28.83	559.1	287.0
4/20/2008	1200	20.91	380.02	1,678.1	3,735.5	5/30/2008	1200	13.54	22.36	508.7	243.7
4/21/2008	1200	19.81	337.56	1,448.5	4,013.6	5/31/2008	1200	11.47	22.73	477.5	229.2
4/22/2008	1200	19.42	270.76	1,247.7	3,682.4	6/1/2008	1200	9.45	32.45	460.6	300.4
4/23/2008	1200	18.62	222.40	1,114.8	3,271.7	6/2/2008	1200	8.61	35.74	447.6	364.0
4/24/2008	1200	18.62	186.12	1,017.5	2,913.0	6/3/2008	1200	8.56	31.08	441.7	353.4
4/25/2008	1200	17.75	150.16	938.3	2,525.0	6/4/2008	1200	7.95	27.68	432.4	314.5
4/26/2008	1200	19.70	120.88	876.9	2,066.7	6/5/2008	1200	7.36	25.18	413.5	284.5
4/27/2008	1200	24.73	115.25	831.3	1,826.3	6/6/2008	1200	6.98	48.94	386.4	450.0
4/28/2008	1200	27.22	113.55	793.1	1,749.0	6/7/2008	1200	6.27	74.93	359.5	919.4
4/29/2008	1200	25.79	144.15	763.4	1,880.8	6/8/2008	1200	5.80	54.33	336.9	834.6
4/30/2008	1200	19.94	131.42	761.9	1,914.6	6/9/2008	1200	5.63	36.10	311.4	619.9
5/1/2008	1200	18.20	100.62	782.0	1,506.6	6/10/2008	1200	6.82	26.97	290.2	464.2
5/2/2008	1200	16.87	79.38	778.8	1,137.0	6/11/2008	1200	26.69	28.43	276.5	393.6
5/3/2008	1200	15.76	72.58	775.2	973.0	6/12/2008	1200	14.62	26.06	262.7	332.6
5/4/2008	1200	15.27	72.40	773.0	964.2	6/13/2008	1200	11.34	19.82	247.9	270.8
5/5/2008	1200	14.46	72.85	761.2	1,024.6	6/14/2008	1200	9.26	15.83	236.8	227.2
5/6/2008	1200	14.04	64.19	748.0	947.1	6/15/2008	1200	8.11	13.97	232.5	197.5
5/7/2008	1200	13.53	54.36	731.7	871.5	6/16/2008	1200	8.07	12.25	235.6	177.5
5/8/2008	1200	12.74	56.48	691.9	839.7	6/17/2008	1200	7.22	10.95	234.6	157.9
5/9/2008	1200	18.86	55.42	663.1	897.0	6/18/2008	1200	5.79	9.81	231.8	142.6
5/10/2008	1200	14.63	54.80	624.8	799.0	6/19/2008	1200	5.14	10.14	232.7	139.9
5/11/2008	1200	71.40	70.17	602.5	666.6	6/20/2008	1200	4.85	10.58	230.6	139.9
5/12/2008	1200	94.41	51.94	644.7	589.1	6/21/2008	1200	5.14	11.65	250.7	146.1

Appendix 1-C. Daily mean streamflow for selected sites from October 1, 2004, to October 1, 2009, used as input into the load estimation model at each site.—Continued

[ft³/s, cubic feet per second; Bolded values were estimated using MOVE.1 regression technique]

Date	Time	Daily mean streamflow (ft³/s)				Date	Time	Daily mean streamflow (ft³/s)			
		McTier Creek 02172305	Fishing Brook 0131199050	Edisto River 02175000	Hudson River 01312000			McTier Creek 02172305	Fishing Brook 0131199050	Edisto River 02175000	Hudson River 01312000
6/22/2008	1200	5.59	14.60	239.7	156.0	8/1/2008	1200	9.29	40.32	298.2	1,085.9
6/23/2008	1200	5.62	17.20	249.9	158.6	8/2/2008	1200	7.52	33.07	287.3	918.7
6/24/2008	1200	5.38	16.41	288.6	175.7	8/3/2008	1200	7.35	34.43	284.6	908.6
6/25/2008	1200	4.77	13.84	373.2	178.0	8/4/2008	1200	6.00	32.67	291.0	886.7
6/26/2008	1200	4.19	12.98	528.4	165.3	8/5/2008	1200	5.24	27.26	284.1	754.8
6/27/2008	1200	4.63	13.84	602.0	159.2	8/6/2008	1200	4.82	135.25	298.1	1,038.2
6/28/2008	1200	4.77	16.91	574.0	162.0	8/7/2008	1200	5.23	256.94	330.5	1,926.6
6/29/2008	1200	4.09	39.53	514.9	264.7	8/8/2008	1200	4.83	142.06	306.9	1,733.2
6/30/2008	1200	4.06	41.96	449.7	425.5	8/9/2008	1200	4.00	108.41	279.7	1,439.9
7/1/2008	1200	4.09	28.61	383.3	452.5	8/10/2008	1200	3.53	87.71	260.9	1,207.1
7/2/2008	1200	3.35	19.91	328.9	344.7	8/11/2008	1200	4.36	85.98	252.2	1,096.0
7/3/2008	1200	2.61	15.97	292.2	272.5	8/12/2008	1200	3.19	87.77	244.9	1,059.2
7/4/2008	1200	2.97	15.88	263.7	247.1	8/13/2008	1200	38.63	68.21	246.5	938.6
7/5/2008	1200	5.80	15.31	240.6	212.3	8/14/2008	1200	30.06	51.87	246.5	816.7
7/6/2008	1200	7.35	12.56	226.1	180.0	8/15/2008	1200	16.81	39.27	238.2	647.4
7/7/2008	1200	10.00	10.02	219.0	154.3	8/16/2008	1200	12.47	31.38	244.7	495.1
7/8/2008	1200	8.12	7.90	233.2	135.1	8/17/2008	1200	10.15	25.55	279.9	392.3
7/9/2008	1200	7.05	8.08	274.8	125.8	8/18/2008	1200	9.54	22.72	314.4	324.6
7/10/2008	1200	14.22	8.00	353.8	129.7	8/19/2008	1200	8.95	50.24	331.6	431.0
7/11/2008	1200	19.64	7.57	452.1	124.9	8/20/2008	1200	8.17	92.21	358.1	653.5
7/12/2008	1200	11.40	11.57	479.6	113.6	8/21/2008	1200	8.59	66.11	390.2	538.8
7/13/2008	1200	8.64	13.19	459.1	117.9	8/22/2008	1200	9.00	42.84	425.1	413.8
7/14/2008	1200	7.35	23.71	434.5	386.4	8/23/2008	1200	17.06	30.09	450.4	325.5
7/15/2008	1200	6.85	26.64	459.9	527.1	8/24/2008	1200	14.59	22.75	455.1	269.9
7/16/2008	1200	6.60	18.66	462.3	353.5	8/25/2008	1200	11.67	20.56	430.5	253.9
7/17/2008	1200	4.71	12.83	448.1	256.4	8/26/2008	1200	11.28	17.67	423.3	227.5
7/18/2008	1200	3.90	9.10	447.0	203.6	8/27/2008	1200	30.96	15.56	420.3	194.4
7/19/2008	1200	3.86	11.00	439.8	174.8	8/28/2008	1200	19.28	13.95	422.8	168.4
7/20/2008	1200	3.95	22.08	424.1	297.3	8/29/2008	1200	14.74	12.22	461.0	148.5
7/21/2008	1200	5.19	97.95	399.0	1,640.8	8/30/2008	1200	12.18	15.89	537.7	136.2
7/22/2008	1200	10.95	97.65	368.0	1,937.8	8/31/2008	1200	10.15	45.74	564.4	129.2
7/23/2008	1200	7.45	66.16	333.3	1,382.5	9/1/2008	1200	8.83	45.44	597.6	121.0
7/24/2008	1200	6.40	72.67	295.4	1,453.2	9/2/2008	1200	7.98	32.93	627.8	114.1
7/25/2008	1200	5.74	98.92	271.5	1,707.9	9/3/2008	1200	7.04	23.50	632.7	106.8
7/26/2008	1200	6.39	80.65	254.3	1,514.8	9/4/2008	1200	6.26	15.81	615.0	99.9
7/27/2008	1200	9.55	96.38	290.0	1,275.0	9/5/2008	1200	6.03	12.40	576.8	92.9
7/28/2008	1200	7.47	102.49	295.3	1,135.9	9/6/2008	1200	6.54	12.28	557.4	89.4
7/29/2008	1200	6.38	67.88	287.4	910.9	9/7/2008	1200	5.97	15.34	497.8	93.9
7/30/2008	1200	5.92	44.02	278.6	699.9	9/8/2008	1200	7.75	16.01	439.3	87.1
7/31/2008	1200	8.28	43.19	282.2	819.2	9/9/2008	1200	7.34	16.55	394.8	91.3

Appendix 1-C.   Daily mean streamflow for selected sites from October 1, 2004, to October 1, 2009, used as input into the load estimation model at each site.—Continued

[ft³/s, cubic feet per second; Bolded values were estimated using MOVE.1 regression technique]

Date	Time	Daily mean streamflow (ft³/s)				Date	Time	Daily mean streamflow (ft³/s)			
		McTier Creek 02172305	Fishing Brook 0131199050	Edisto River 02175000	Hudson River 01312000			McTier Creek 02172305	Fishing Brook 0131199050	Edisto River 02175000	Hudson River 01312000
9/10/2008	1200	14.57	19.44	361.0	99.9	10/21/2008	1200	12.79	23.41	529.2	142.6
9/11/2008	1200	15.02	18.35	336.1	93.1	10/22/2008	1200	12.00	24.53	501.5	156.3
9/12/2008	1200	16.93	16.19	348.4	87.9	10/23/2008	1200	11.68	24.94	477.0	157.0
9/13/2008	1200	14.25	20.50	371.0	106.9	10/24/2008	1200	20.97	23.80	548.2	146.6
9/14/2008	1200	11.48	37.76	434.0	197.4	10/25/2008	1200	47.64	39.23	1,240.7	157.6
9/15/2008	1200	9.69	60.44	455.8	472.7	10/26/2008	1200	28.32	268.84	2,216.1	1,029.5
9/16/2008	1200	9.08	46.32	460.5	410.1	10/27/2008	1200	19.85	219.61	2,409.7	1,822.0
9/17/2008	1200	9.52	31.73	481.9	306.9	10/28/2008	1200	16.18	129.22	2,112.1	1,437.1
9/18/2008	1200	9.65	22.84	591.8	244.3	10/29/2008	1200	17.31	118.59	1,857.4	1,136.1
9/19/2008	1200	8.63	16.80	531.0	197.0	10/30/2008	1200	15.04	96.21	1,753.5	936.3
9/20/2008	1200	8.11	13.33	422.1	165.0	10/31/2008	1200	13.63	77.40	1,711.1	771.4
9/21/2008	1200	7.58	10.09	383.1	143.7	11/1/2008	1200	13.26	82.73	1,644.9	677.6
9/22/2008	1200	7.80	8.79	360.0	128.1	11/2/2008	1200	13.03	94.79	1,555.4	705.0
9/23/2008	1200	6.98	8.23	334.6	114.5	11/3/2008	1200	13.15	83.39	1,477.0	645.0
9/24/2008	1200	6.43	7.23	310.0	103.8	11/4/2008	1200	14.28	71.06	1,406.0	557.9
9/25/2008	1200	5.72	6.33	294.7	94.4	11/5/2008	1200	15.61	81.76	1,331.9	516.3
9/26/2008	1200	7.82	6.47	302.1	89.2	11/6/2008	1200	15.65	97.45	1,237.0	564.4
9/27/2008	1200	12.04	8.56	295.2	94.2	11/7/2008	1200	15.06	86.55	1,132.9	580.7
9/28/2008	1200	10.70	9.97	288.4	97.5	11/8/2008	1200	14.34	73.30	1,031.6	552.9
9/29/2008	1200	9.82	12.34	288.3	95.1	11/9/2008	1200	13.58	67.56	932.9	546.4
9/30/2008	1200	20.33	14.51	301.2	93.5	11/10/2008	1200	14.06	64.86	841.7	537.1
10/1/2008	1200	15.56	14.85	308.2	106.2	11/11/2008	1200	14.96	58.62	760.6	486.4
10/2/2008	1200	10.82	16.50	310.2	150.0	11/12/2008	1200	14.56	50.94	697.0	422.7
10/3/2008	1200	8.97	23.98	307.4	185.7	11/13/2008	1200	24.79	46.52	655.1	369.1
10/4/2008	1200	8.13	34.04	303.3	212.8	11/14/2008	1200	113.14	53.02	625.0	357.9
10/5/2008	1200	7.79	34.38	294.1	208.2	11/15/2008	1200	154.00	74.75	654.6	506.7
10/6/2008	1200	7.39	27.59	291.1	188.8	11/16/2008	1200	49.00	149.92	741.1	1,068.9
10/7/2008	1200	7.44	21.88	286.9	168.9	11/17/2008	1200	31.00	145.40	851.4	1,416.0
10/8/2008	1200	10.43	18.25	289.7	152.1	11/18/2008	1200	22.73	100.77	958.2	1,177.4
10/9/2008	1200	72.10	20.44	328.2	171.9	11/19/2008	1200	20.44	73.24	1,068.0	917.5
10/10/2008	1200	33.71	21.31	406.2	278.3	11/20/2008	1200	21.00	57.08	1,143.3	712.5
10/11/2008	1200	24.08	21.41	617.9	264.2	11/21/2008	1200	21.00	47.67	1,194.5	532.1
10/12/2008	1200	18.68	19.73	707.6	220.6	11/22/2008	1200	16.12	41.54	1,248.2	411.2
10/13/2008	1200	15.79	17.69	750.4	187.8	11/23/2008	1200	15.00	36.87	1,311.3	332.8
10/14/2008	1200	13.56	16.84	768.2	163.0	11/24/2008	1200	16.34	34.48	1,388.0	295.6
10/15/2008	1200	12.58	14.76	751.8	144.5	11/25/2008	1200	17.11	37.79	1,461.6	284.5
10/16/2008	1200	11.83	19.39	701.1	141.9	11/26/2008	1200	16.00	39.33	1,527.8	285.7
10/17/2008	1200	12.06	28.68	649.2	164.1	11/27/2008	1200	16.18	36.88	1,596.3	272.3
10/18/2008	1200	16.18	32.17	610.4	168.9	11/28/2008	1200	16.08	34.98	1,669.5	253.8
10/19/2008	1200	14.73	27.77	583.7	156.7	11/29/2008	1200	41.51	33.93	1,735.1	239.5
10/20/2008	1200	13.04	23.65	558.8	144.8	11/30/2008	1200	122.00	32.53	1,850.9	223.2

Appendix 1-C. Daily mean streamflow for selected sites from October 1, 2004, to October 1, 2009, used as input into the load estimation model at each site.—Continued

[ft³/s, cubic feet per second; Bolded values were estimated using MOVE.1 regression technique]

Date	Time	Daily mean streamflow (ft³/s)				Date	Time	Daily mean streamflow (ft³/s)			
		McTier Creek 02172305	Fishing Brook 0131199050	Edisto River 02175000	Hudson River 01312000			McTier Creek 02172305	Fishing Brook 0131199050	Edisto River 02175000	Hudson River 01312000
12/1/2008	1200	131.00	43.40	2,114.0	246.8	1/11/2009	1200	33.76	36.92	1,499.4	257.2
12/2/2008	1200	56.10	56.07	2,387.5	292.7	1/12/2009	1200	31.02	34.34	1,482.9	241.1
12/3/2008	1200	35.50	52.65	2,505.6	293.3	1/13/2009	1200	28.15	33.04	1,471.9	226.3
12/4/2008	1200	29.61	45.24	2,459.8	274.8	1/14/2009	1200	29.33	31.10	1,445.6	215.3
12/5/2008	1200	28.66	41.66	2,442.3	256.4	1/15/2009	1200	27.81	29.71	1,417.6	203.9
12/6/2008	1200	25.94	36.56	2,516.2	225.8	1/16/2009	1200	27.81	27.89	1,396.0	195.2
12/7/2008	1200	24.44	34.69	2,673.1	219.5	1/17/2009	1200	23.75	26.48	1,396.6	187.5
12/8/2008	1200	22.30	29.25	2,849.7	190.7	1/18/2009	1200	25.79	26.27	1,405.9	181.9
12/9/2008	1200	21.64	27.18	2,987.2	181.1	1/19/2009	1200	32.92	26.44	1,433.1	177.4
12/10/2008	1200	25.94	56.73	3,043.3	224.4	1/20/2009	1200	28.21	26.37	1,452.8	171.5
12/11/2008	1200	125.19	106.37	3,024.2	468.3	1/21/2009	1200	27.17	25.60	1,464.3	164.8
12/12/2008	1200	162.59	103.01	2,939.8	685.2	1/22/2009	1200	30.63	24.84	1,471.8	160.9
12/13/2008	1200	114.83	76.40	2,815.6	547.7	1/23/2009	1200	26.13	24.07	1,483.4	156.9
12/14/2008	1200	53.66	58.07	2,696.6	412.8	1/24/2009	1200	25.09	24.27	1,500.1	152.8
12/15/2008	1200	42.75	59.72	2,592.0	382.2	1/25/2009	1200	25.33	23.61	1,510.5	147.3
12/16/2008	1200	38.41	107.83	2,504.3	468.9	1/26/2009	1200	23.48	22.79	1,494.3	144.7
12/17/2008	1200	35.34	109.52	2,471.8	640.3	1/27/2009	1200	25.41	22.09	1,471.9	141.1
12/18/2008	1200	33.44	85.56	2,510.8	585.1	1/28/2009	1200	25.77	22.98	1,458.1	140.7
12/19/2008	1200	31.12	68.02	2,584.8	495.2	1/29/2009	1200	26.44	23.53	1,486.9	149.1
12/20/2008	1200	30.88	57.71	2,668.0	407.1	1/30/2009	1200	23.49	23.55	1,509.5	149.3
12/21/2008	1200	30.16	51.41	2,696.2	368.3	1/31/2009	1200	21.50	22.69	1,505.1	146.9
12/22/2008	1200	26.98	42.99	2,662.9	332.3	2/1/2009	1200	23.51	22.06	1,483.3	142.9
12/23/2008	1200	26.61	42.95	2,658.8	305.8	2/2/2009	1200	25.73	21.41	1,463.0	137.9
12/24/2008	1200	29.98	40.88	2,728.4	294.3	2/3/2009	1200	24.81	21.00	1,441.1	133.0
12/25/2008	1200	27.36	58.93	2,752.9	322.4	2/4/2009	1200	21.92	20.24	1,414.9	128.4
12/26/2008	1200	28.01	90.78	2,675.6	416.1	2/5/2009	1200	19.98	19.19	1,382.7	124.9
12/27/2008	1200	29.49	81.95	2,515.8	478.8	2/6/2009	1200	19.86	18.94	1,341.3	122.3
12/28/2008	1200	28.32	148.18	2,322.6	671.5	2/7/2009	1200	20.62	19.04	1,300.9	120.3
12/29/2008	1200	27.20	260.06	2,127.7	1,201.8	2/8/2009	1200	19.90	23.04	1,256.2	124.8
12/30/2008	1200	27.24	180.35	1,953.4	1,338.8	2/9/2009	1200	19.85	27.62	1,218.0	128.2
12/31/2008	1200	25.24	115.30	1,842.7	1,119.6	2/10/2009	1200	19.75	24.90	1,179.7	128.9
1/1/2009	1200	22.04	82.33	1,775.9	884.1	2/11/2009	1200	19.96	24.52	1,140.5	128.4
1/2/2009	1200	25.98	64.39	1,718.3	688.7	2/12/2009	1200	20.32	47.56	1,112.7	150.1
1/3/2009	1200	24.67	59.19	1,655.9	560.3	2/13/2009	1200	18.54	74.66	1,083.9	205.5
1/4/2009	1200	25.01	50.59	1,600.5	489.4	2/14/2009	1200	26.48	68.87	1,060.3	291.5
1/5/2009	1200	26.18	47.13	1,557.3	411.2	2/15/2009	1200	30.12	50.30	1,037.3	333.9
1/6/2009	1200	26.54	43.24	1,528.4	354.9	2/16/2009	1200	23.24	39.73	1,021.7	308.9
1/7/2009	1200	35.72	44.04	1,519.2	326.2	2/17/2009	1200	20.98	33.80	1,034.7	268.7
1/8/2009	1200	34.15	41.08	1,518.5	310.3	2/18/2009	1200	28.15	30.83	1,085.4	235.9
1/9/2009	1200	26.08	38.25	1,518.8	286.8	2/19/2009	1200	121.23	30.01	1,177.3	218.1
1/10/2009	1200	24.49	36.29	1,512.0	271.2	2/20/2009	1200	64.49	27.69	1,288.5	200.8

Appendix 1-C.   Daily mean streamflow for selected sites from October 1, 2004, to October 1, 2009, used as input into the load estimation model at each site.—Continued

[ft³/s, cubic feet per second; Bolded values were estimated using MOVE.1 regression technique]

Date	Time	McTier Creek 02172305	Fishing Brook 0131199050	Edisto River 02175000	Hudson River 01312000	Date	Time	McTier Creek 02172305	Fishing Brook 0131199050	Edisto River 02175000	Hudson River 01312000
2/21/2009	1200	39.09	26.28	1,378.0	185.6	4/2/2009	1200	57.01	123.47	2,856.6	1,406.0
2/22/2009	1200	33.12	27.23	1,446.1	177.9	4/3/2009	1200	82.37	161.15	3,590.7	1,436.4
2/23/2009	1200	26.55	25.54	1,567.0	170.9	4/4/2009	1200	51.95	310.06	4,392.4	2,309.6
2/24/2009	1200	24.05	23.55	1,715.5	161.5	4/5/2009	1200	38.06	233.32	4,743.4	2,675.5
2/25/2009	1200	22.56	23.31	1,848.6	152.0	4/6/2009	1200	35.09	145.77	4,712.3	2,009.3
2/26/2009	1200	22.24	22.81	1,938.0	144.9	4/7/2009	1200	31.74	126.42	4,596.5	1,558.8
2/27/2009	1200	24.56	31.94	1,978.9	148.0	4/8/2009	1200	30.03	111.63	4,503.4	1,328.4
2/28/2009	1200	97.19	78.51	1,991.0	199.8	4/9/2009	1200	27.46	95.08	4,446.5	1,103.5
3/1/2009	1200	158.89	92.11	2,128.0	303.7	4/10/2009	1200	25.55	83.96	4,345.7	951.6
3/2/2009	1200	129.92	69.98	2,508.8	401.3	4/11/2009	1200	104.52	79.89	4,193.6	879.0
3/3/2009	1200	63.77	48.25	2,805.3	366.5	4/12/2009	1200	80.85	75.68	3,956.2	848.7
3/4/2009	1200	46.82	37.40	2,910.1	305.1	4/13/2009	1200	44.28	67.98	3,667.1	772.8
3/5/2009	1200	41.19	32.35	2,911.4	260.0	4/14/2009	1200	54.65	62.42	3,386.5	676.2
3/6/2009	1200	37.91	30.79	2,925.4	230.2	4/15/2009	1200	73.04	60.51	3,155.8	632.4
3/7/2009	1200	35.43	38.96	3,016.5	214.7	4/16/2009	1200	44.89	61.77	2,965.0	672.8
3/8/2009	1200	33.41	77.85	3,164.0	266.6	4/17/2009	1200	36.46	64.96	2,879.8	740.7
3/9/2009	1200	31.18	112.81	3,343.1	430.0	4/18/2009	1200	31.93	71.49	2,856.8	832.2
3/10/2009	1200	28.96	100.75	3,497.7	594.5	4/19/2009	1200	29.24	91.60	2,855.0	936.0
3/11/2009	1200	28.84	94.35	3,581.1	642.7	4/20/2009	1200	36.62	92.05	2,845.4	957.7
3/12/2009	1200	28.47	119.03	3,616.0	739.8	4/21/2009	1200	32.89	100.66	2,818.6	947.5
3/13/2009	1200	26.81	108.75	3,555.4	792.7	4/22/2009	1200	28.49	122.97	2,760.7	1,031.9
3/14/2009	1200	27.49	82.30	3,376.7	719.7	4/23/2009	1200	25.44	123.07	2,683.7	1,177.3
3/15/2009	1200	37.83	65.73	3,090.6	580.8	4/24/2009	1200	23.49	104.95	2,571.2	1,118.6
3/16/2009	1200	54.92	60.47	2,802.8	486.0	4/25/2009	1200	22.06	103.62	2,425.8	1,089.0
3/17/2009	1200	50.30	66.16	2,644.1	450.3	4/26/2009	1200	20.28	142.61	2,233.1	1,466.2
3/18/2009	1200	36.31	86.53	2,600.8	499.6	4/27/2009	1200	20.22	132.32	2,010.5	1,724.7
3/19/2009	1200	30.55	118.61	2,564.3	630.0	4/28/2009	1200	19.43	113.83	1,788.0	1,668.5
3/20/2009	1200	27.13	123.90	2,528.1	775.5	4/29/2009	1200	18.99	98.34	1,562.4	1,604.9
3/21/2009	1200	25.33	100.15	2,511.7	787.7	4/30/2009	1200	18.77	77.80	1,361.2	1,300.0
3/22/2009	1200	26.13	85.11	2,489.8	721.1	5/1/2009	1200	17.82	79.41	1,190.7	1,080.4
3/23/2009	1200	23.79	64.26	2,466.7	579.8	5/2/2009	1200	18.37	105.14	1,052.6	1,437.4
3/24/2009	1200	23.66	55.73	2,422.2	499.1	5/3/2009	1200	20.23	98.90	943.5	1,410.8
3/25/2009	1200	24.68	54.18	2,347.8	433.7	5/4/2009	1200	19.31	76.78	872.2	1,100.3
3/26/2009	1200	25.30	56.39	2,247.5	402.0	5/5/2009	1200	32.97	60.39	969.8	880.1
3/27/2009	1200	30.46	81.48	2,148.2	467.2	5/6/2009	1200	34.97	64.54	1,016.5	786.8
3/28/2009	1200	59.13	121.12	2,059.4	753.2	5/7/2009	1200	34.09	91.95	1,003.6	850.4
3/29/2009	1200	133.80	189.64	2,,174.9	1,130.5	5/8/2009	1200	29.78	108.81	1,055.7	1,074.6
3/30/2009	1200	60.02	352.97	2,426.9	1,873.8	5/9/2009	1200	22.20	104.38	1,168.8	1,100.9
3/31/2009	1200	38.59	237.33	2,575.1	2,144.4	5/10/2009	1200	19.88	172.52	1,344.4	1,378.9
4/1/2009	1200	33.38	148.41	2,640.5	1,748.0	5/11/2009	1200	20.25	165.14	1,890.9	1,534.7

Appendix 1-C. Daily mean streamflow for selected sites from October 1, 2004, to October 1, 2009, used as input into the load estimation model at each site.—Continued

[ft³/s, cubic feet per second; Bolded values were estimated using MOVE.1 regression technique]

Date	Time	Daily mean streamflow (ft³/s)				Date	Time	Daily mean streamflow (ft³/s)			
		McTier Creek 02172305	Fishing Brook 0131199050	Edisto River 02175000	Hudson River 01312000			McTier Creek 02172305	Fishing Brook 0131199050	Edisto River 02175000	Hudson River 01312000
5/12/2009	1200	21.15	121.60	2,512.8	1,263.0	6/21/2009	1200	8.92	37.34	1,040.4	650.0
5/13/2009	1200	18.39	92.74	2,473.7	1,016.8	6/22/2009	1200	9.47	31.22	946.9	498.4
5/14/2009	1200	17.04	75.79	2,170.1	853.3	6/23/2009	1200	11.94	25.82	853.3	424.2
5/15/2009	1200	16.97	76.47	1,914.9	854.5	6/24/2009	1200	9.16	20.91	768.9	330.8
5/16/2009	1200	27.84	81.45	1,790.1	860.1	6/25/2009	1200	7.62	17.03	726.7	267.4
5/17/2009	1200	45.58	235.31	1,756.7	1,623.8	6/26/2009	1200	6.99	14.68	696.3	224.4
5/18/2009	1200	54.56	202.91	2,006.2	2,133.1	6/27/2009	1200	6.71	18.87	651.3	215.4
5/19/2009	1200	32.42	126.55	2,745.0	1,606.5	6/28/2009	1200	5.98	30.02	653.0	239.8
5/20/2009	1200	22.94	92.72	2,893.1	1,156.2	6/29/2009	1200	5.60	32.99	672.8	247.3
5/21/2009	1200	24.71	72.13	2,616.7	914.9	6/30/2009	1200	5.06	27.81	615.8	235.3
5/22/2009	1200	22.13	58.68	2,508.8	744.2	7/1/2009	1200	4.46	25.44	540.9	241.3
5/23/2009	1200	20.51	48.11	2,518.3	576.9	7/2/2009	1200	3.92	21.94	477.6	249.9
5/24/2009	1200	23.00	44.45	2,599.3	464.6	7/3/2009	1200	3.70	23.14	423.9	240.1
5/25/2009	1200	30.00	42.18	2,865.7	388.1	7/4/2009	1200	3.83	36.45	384.9	338.0
5/26/2009	1200	26.00	35.05	3,359.1	318.5	7/5/2009	1200	4.39	55.17	351.1	758.5
5/27/2009	1200	23.06	39.38	3,703.4	319.0	7/6/2009	1200	9.29	65.73	352.0	721.8
5/28/2009	1200	30.92	90.73	3,630.9	797.3	7/7/2009	1200	8.40	53.57	393.2	528.2
5/29/2009	1200	29.90	127.04	3,459.4	1,645.1	7/8/2009	1200	43.73	53.11	442.5	566.5
5/30/2009	1200	21.24	112.20	3,232.9	1,835.9	7/9/2009	1200	68.51	54.06	750.9	703.1
5/31/2009	1200	19.70	89.28	3,099.8	1,499.5	7/10/2009	1200	25.25	44.44	1,174.1	621.3
6/1/2009	1200	15.86	65.68	3,026.9	1,113.1	7/11/2009	1200	15.74	36.63	1,500.4	492.1
6/2/2009	1200	12.67	53.40	2,825.4	876.5	7/12/2009	1200	11.61	53.98	1,698.5	615.3
6/3/2009	1200	11.44	44.61	2,551.8	691.0	7/13/2009	1200	13.93	53.48	1,736.8	761.7
6/4/2009	1200	17.09	38.84	2,246.3	520.5	7/14/2009	1200	15.10	42.08	1,670.0	617.0
6/5/2009	1200	50.57	37.78	1,941.4	405.2	7/15/2009	1200	11.53	33.84	1,744.7	495.0
6/6/2009	1200	34.44	33.48	1,763.5	330.2	7/16/2009	1200	10.73	26.92	1,904.7	409.4
6/7/2009	1200	23.47	27.57	1,698.6	280.9	7/17/2009	1200	10.33	21.03	1,889.9	344.4
6/8/2009	1200	19.81	24.34	1,735.6	246.5	7/18/2009	1200	10.56	17.87	1,780.6	315.2
6/9/2009	1200	16.60	25.26	1,790.2	246.3	7/19/2009	1200	8.71	14.62	1,603.4	290.8
6/10/2009	1200	14.31	26.24	1,753.0	264.8	7/20/2009	1200	7.85	12.78	1,413.3	255.7
6/11/2009	1200	13.32	23.74	1,706.9	245.2	7/21/2009	1200	7.72	11.29	1,307.8	223.7
6/12/2009	1200	14.57	43.50	1,627.2	315.1	7/22/2009	1200	7.60	10.82	1,256.9	205.0
6/13/2009	1200	13.27	54.45	1,504.3	460.9	7/23/2009	1200	8.72	9.99	1,173.0	186.8
6/14/2009	1200	11.95	48.02	1,467.0	410.5	7/24/2009	1200	7.64	10.38	1,057.3	171.1
6/15/2009	1200	10.98	52.97	1,533.1	370.9	7/25/2009	1200	7.88	11.25	955.8	166.7
6/16/2009	1200	10.29	51.19	1,597.4	585.1	7/26/2009	1200	6.74	13.28	870.5	159.2
6/17/2009	1200	11.09	39.52	1,577.2	602.5	7/27/2009	1200	5.83	14.19	795.7	149.6
6/18/2009	1200	12.24	32.65	1,453.2	458.5	7/28/2009	1200	5.75	14.16	736.3	138.2
6/19/2009	1200	11.09	39.48	1,297.7	637.6	7/29/2009	1200	6.22	14.13	713.2	130.3
6/20/2009	1200	9.85	42.13	1,139.2	805.2	7/30/2009	1200	9.09	26.29	694.7	162.7

Appendix 1-C. Daily mean streamflow for selected sites from October 1, 2004, to October 1, 2009, used as input into the load estimation model at each site.—Continued

[ft³/s, cubic feet per second; Bolded values were estimated using MOVE.1 regression technique]

Date	Time	Daily mean streamflow (ft³/s)				Date	Time	Daily mean streamflow (ft³/s)			
		McTier Creek 02172305	Fishing Brook 0131199050	Edisto River 02175000	Hudson River 01312000			McTier Creek 02172305	Fishing Brook 0131199050	Edisto River 02175000	Hudson River 01312000
7/31/2009	1200	20.15	41.03	627.9	215.0	9/9/2009	1200	7.28	5.87	373.3	88.8
8/1/2009	1200	23.43	50.04	555.5	327.2	9/10/2009	1200	8.65	5.45	358.7	80.7
8/2/2009	1200	16.00	50.31	521.4	340.8	9/11/2009	1200	10.39	5.16	347.4	71.7
8/3/2009	1200	15.49	74.75	517.0	463.3	9/12/2009	1200	9.10	5.56	336.2	66.6
8/4/2009	1200	11.86	57.04	509.1	507.2	9/13/2009	1200	8.75	6.27	328.7	65.0
8/5/2009	1200	10.00	37.46	491.3	406.8	9/14/2009	1200	7.90	6.26	324.9	61.7
8/6/2009	1200	8.77	25.90	483.6	320.3	9/15/2009	1200	7.59	6.49	323.2	61.8
8/7/2009	1200	7.79	18.61	508.4	259.4	9/16/2009	1200	7.59	10.21	326.3	59.1
8/8/2009	1200	7.04	14.15	513.2	213.6	9/17/2009	1200	7.10	12.89	329.9	55.5
8/9/2009	1200	6.80	11.68	500.1	180.9	9/18/2009	1200	12.02	12.66	330.8	53.5
8/10/2009	1200	7.19	11.17	468.5	162.2	9/19/2009	1200	13.64	9.68	335.2	53.5
8/11/2009	1200	5.44	11.08	432.8	155.1	9/20/2009	1200	12.87	7.67	325.5	51.7
8/12/2009	1200	9.28	9.57	418.1	143.7	9/21/2009	1200	20.59	6.55	340.9	48.5
8/13/2009	1200	71.92	8.66	404.8	194.3	9/22/2009	1200	13.87	7.19	365.2	48.7
8/14/2009	1200	86.32	8.50	414.8	223.8	9/23/2009	1200	10.91	8.90	384.5	52.1
8/15/2009	1200	36.06	7.69	475.0	201.5	9/24/2009	1200	10.65	11.23	394.8	62.8
8/16/2009	1200	21.48	7.19	579.7	160.4	9/25/2009	1200	9.95	11.49	401.1	61.2
8/17/2009	1200	20.66	6.54	616.4	132.2	9/26/2009	1200	9.78	10.36	431.1	57.3
8/18/2009	1200	14.45	6.52	579.7	115.0	9/27/2009	1200	16.78	15.33	465.4	71.6
8/19/2009	1200	11.14	11.82	555.8	117.5	9/28/2009	1200	13.26	42.53	484.4	282.3
8/20/2009	1200	12.20	15.78	580.8	126.8	9/29/2009	1200	10.13	63.42	484.5	450.0
8/21/2009	1200	15.78	18.35	609.8	188.7	9/30/2009	1200	9.04	56.92	460.8	507.2
8/22/2009	1200	10.78	19.50	610.2	181.2	10/01/09	1200	8.55	56.00	452.6	507.0
8/23/2009	1200	8.61	33.92	590.4	172.8						
8/24/2009	1200	7.34	70.83	550.4	199.3						
8/25/2009	1200	8.02	52.61	500.0	236.5						
8/26/2009	1200	7.14	33.37	459.9	216.1						
8/27/2009	1200	7.02	23.92	446.8	188.0						
8/28/2009	1200	8.15	17.95	444.9	161.4						
8/29/2009	1200	8.77	22.92	429.9	193.9						
8/30/2009	1200	15.18	31.70	403.8	344.8						
8/31/2009	1200	44.37	31.16	390.5	387.2						
9/1/2009	1200	24.45	24.11	377.6	302.8						
9/2/2009	1200	19.21	18.74	376.2	239.4						
9/3/2009	1200	14.24	14.87	376.3	196.3						
9/4/2009	1200	10.74	12.20	378.2	165.4						
9/5/2009	1200	8.94	10.25	378.7	140.8						
9/6/2009	1200	8.23	7.98	380.4	122.6						
9/7/2009	1200	7.33	7.09	379.4	108.0						
9/8/2009	1200	7.19	6.61	382.4	97.4						

Appendix 3-A. Calculated loads of fluvial total mercury (THg) and methylmercury (MeHg) for water years 2005 to 2009, with lower and upper bounds of the 95 percent confidence interval (CI) for the model.

[kg/yr, kilograms per year; (μg/m^2)/yr, micrograms per square meter per year. Total mercury includes methylated and inorganic forms of mercury. A water year extends from October of one calendar year to September of the following calendar year. For the McTier Creek and Fishing Brook sites, discharge records were extended to water years 2005, 2006, and 2007 using MOVE 1 estimation techniques; therefore, the 2005, 2006, and 2007 annual loads at these two sites were estimated using the extended discharge record. Gaged discharge data were used for load computations for all annual loads at Edisto River and Hudson River sites and for 2008 and 2009 annual loads for the McTier Creek and Fishing Brook sites]

	A. Total mercury (THg) loads, by water year								
	Water year 2005								
Site	**Filtered THg annual loads (kg/yr)**			**Particulate THg annual loads (kg/yr)**			**Filtered plus particulate THg annual loads (kg/yr)**		
	Load	Lower CI	Upper CI	Load	Lower CI	Upper CI	Load	Lower CI	Upper CI
McTier Creek, S.C.	0.108	0.084	0.136	0.058	0.037	0.086	0.166	0.121	0.222
Edisto River, S.C.	7.834	5.360	11.060	2.861	1.327	5.424	10.695	6.687	16.484
Fishing Brook, N.Y.	0.084	0.077	0.091	0.016	0.013	0.018	0.099	0.090	0.109
Hudson River, N.Y.	0.600	0.563	0.640	0.156	0.121	0.199	0.757	0.684	0.839
	Water year 2006								
McTier Creek, S.C.	0.081	0.067	0.097	0.038	0.028	0.049	0.119	0.095	0.146
Edisto River, S.C.	4.007	3.205	4.949	1.828	1.044	2.978	5.835	4.249	7.927
Fishing Brook, N.Y.	0.127	0.115	0.139	0.023	0.020	0.027	0.150	0.135	0.166
Hudson River, N.Y.	0.921	0.866	0.980	0.244	0.195	0.302	1.165	1.061	1.282
	Water year 2007								
McTier Creek, S.C.	0.079	0.060	0.101	0.030	0.021	0.042	0.109	0.081	0.143
Edisto River, S.C.	5.146	3.847	6.742	1.693	0.990	2.710	6.839	4.837	9.452
Fishing Brook, N.Y.	0.107	0.097	0.117	0.020	0.017	0.023	0.127	0.114	0.140
Hudson River, N.Y.	0.803	0.746	0.863	0.223	0.167	0.291	1.026	0.913	1.154
	Water year 2008								
McTier Creek, S.C.	0.053	0.044	0.063	0.020	0.015	0.025	0.073	0.059	0.088
Edisto River, S.C.	3.026	2.359	3.823	1.381	0.749	2.341	4.407	3.108	6.164
Fishing Brook, N.Y.	0.116	0.106	0.127	0.021	0.018	0.025	0.138	0.124	0.152
Hudson River, N.Y.	0.922	0.859	0.989	0.255	0.192	0.332	1.177	1.051	1.321
	Water year 2009								
McTier Creek, S.C.	0.080	0.063	0.100	0.035	0.025	0.047	0.114	0.088	0.147
Edisto River, S.C.	6.637	4.918	8.763	2.434	1.287	4.203	9.071	6.205	12.966
Fishing Brook, N.Y.	0.102	0.093	0.110	0.018	0.016	0.021	0.120	0.109	0.131
Hudson River, N.Y.	0.733	0.690	0.777	0.187	0.150	0.230	0.920	0.840	1.007

	5-Year mean annual loads (kg/yr)			**2-Year mean annual loads (kg/yr)**			**5-Year lower CI**			**5-Year upper CI**		
	Filtered THg	Particulate THg	THg	Filtered THg	Particulate THg	THg	Filtered THg	Particulate THg	THg	Filtered THg	Particulate THg	THg
McTier Creek, S.C.	0.080	0.036	0.116	0.066	0.027	0.094	0.064	0.025	0.089	0.099	0.050	0.149
Edisto River, S.C.	5.330	2.039	7.369	4.832	1.908	6.739	3.938	1.079	5.017	7.067	3.531	10.599
Fishing Brook, N.Y.	0.107	0.020	0.127	0.109	0.020	0.129	0.098	0.017	0.115	0.117	0.023	0.139
Hudson River, N.Y.	0.796	0.213	1.009	0.827	0.221	1.048	0.745	0.165	0.910	0.850	0.271	1.121

Appendix 3-A.   Calculated loads of fluvial total mercury (THg) and methylmercury (MeHg) for water years 2005 to 2009, with lower and upper bounds of the 95 percent confidence interval (CI) for the model.—Continued

[kg/yr, kilograms per year; (μg/m²)/yr, micrograms per square meter per year. Total mercury includes methylated and inorganic forms of mercury. A water year extends from October of one calendar year to September of the following calendar year. For the McTier Creek and Fishing Brook sites, discharge records were extended to water years 2005, 2006, and 2007 using MOVE 1 estimation techniques; therefore, the 2005, 2006, and 2007 annual loads at these two sites were estimated using the extended discharge record Gaged discharge data were used for load computations for all annual loads at Edisto River and Hudson River sites and for 2008 and 2009 annual loads for the McTier Creek and Fishing Brook sites]

	B. Methylmercury (MeHg) loads, by water year								
	Water year 2005								
Site	Filtered MeHg annual loads (kg/yr)			Particulate MeHg annual loads (kg/yr)			Filtered plus particulate MeHg annual loads (kg/yr)		
	Load	Lower CI	Upper CI	Load	Lower CI	Upper CI	Load	Lower CI	Upper CI
McTier Creek, S.C.	0.003	0.003	0.004	0.002	0.001	0.004	0.005	0.004	0.008
Edisto River, S.C.	0.909	0.543	1.431	0.137	0.076	0.230	1.046	0.618	1.661
Fishing Brook, N.Y.	0.005	0.004	0.006	0.001	0.001	0.001	0.006	0.005	0.007
Hudson River, N.Y.	0.025	0.022	0.029	0.006	0.005	0.008	0.031	0.026	0.037
	Water year 2006								
McTier Creek, S.C.	0.002	0.002	0.003	0.001	0.001	0.002	0.003	0.003	0.004
Edisto River, S.C.	0.375	0.285	0.485	0.071	0.048	0.103	0.446	0.332	0.588
Fishing Brook, N.Y.	0.007	0.006	0.009	0.001	0.001	0.002	0.009	0.007	0.011
Hudson River, N.Y.	0.038	0.032	0.044	0.010	0.008	0.014	0.048	0.040	0.058
	Water year 2007								
McTier Creek, S.C.	0.002	0.002	0.002	0.001	0.000	0.001	0.003	0.002	0.003
Edisto River, S.C.	0.407	0.296	0.547	0.064	0.044	0.090	0.471	0.340	0.636
Fishing Brook, N.Y.	0.005	0.004	0.006	0.001	0.001	0.001	0.006	0.005	0.008
Hudson River, N.Y.	0.027	0.023	0.032	0.008	0.006	0.011	0.035	0.029	0.043
	Water year 2008								
McTier Creek, S.C.	0.002	0.002	0.002	0.001	0.000	0.001	0.002	0.002	0.003
Edisto River, S.C.	0.255	0.185	0.342	0.050	0.033	0.075	0.305	0.218	0.417
Fishing Brook, N.Y.	0.006	0.005	0.008	0.001	0.001	0.001	0.007	0.006	0.009
Hudson River, N.Y.	0.035	0.030	0.041	0.010	0.007	0.014	0.046	0.038	0.056
	Water year 2009								
McTier Creek, S.C.	0.002	0.002	0.003	0.001	0.001	0.001	0.003	0.003	0.004
Edisto River, S.C.	0.644	0.445	0.902	0.106	0.066	0.162	0.750	0.511	1.063
Fishing Brook, N.Y.	0.006	0.005	0.007	0.001	0.001	0.001	0.007	0.006	0.008
Hudson River, N.Y.	0.032	0.027	0.037	0.008	0.006	0.010	0.040	0.033	0.047

	5-Year mean annual loads (kg/yr)			2-Year mean annual loads (kg/yr)			5-Year lower CI			5-Year upper CI		
	Filtered MeHg	Particulate MeHg	MeHg	Filtered MeHg	Particulate MeHg	MeHg	Filtered MeHg	Particulate MeHg	MeHg	Filtered MeHg	Particulate MeHg	MeHg
McTier Creek, S.C.	0.002	0.001	0.003	0.002	0.001	0.003	0.002	0.001	0.003	0.003	0.002	0.004
Edisto River, S.C.	0.518	0.086	0.604	0.449	0.078	0.527	0.351	0.053	0.404	0.741	0.132	0.873
Fishing Brook, N.Y.	0.006	0.001	0.007	0.006	0.001	0.007	0.005	0.001	0.006	0.007	0.001	0.008
Hudson River, N.Y.	0.031	0.009	0.040	0.034	0.009	0.043	0.027	0.006	0.028	0.037	0.011	0.048

Appendix 3-A.   Calculated loads of fluvial total mercury (THg) and methylmercury (MeHg) for water years 2005 to 2009, with lower and upper bounds of the 95 percent confidence interval (CI) for the model.—Continued

[kg/yr, kilograms per year; ($\mu g/m^2$)/yr, micrograms per square meter per year. Total mercury includes methylated and inorganic forms of mercury. A water year extends from October of one calendar year to September of the following calendar year. For the McTier Creek and Fishing Brook sites, discharge records were extended to water years 2005, 2006, and 2007 using MOVE 1 estimation techniques; therefore, the 2005, 2006, and 2007 annual loads at these two sites were estimated using the extended discharge record Gaged discharge data were used for load computations for all annual loads at Edisto River and Hudson River sites and for 2008 and 2009 annual loads for the McTier Creek and Fishing Brook sites]

	C. Total mercury (THg) yields, by water year								
	Water year 2005								
Site	Filtered THg annual yields [($\mu g/m^2$)/yr]			Particulate THg annual yields [($\mu g/m^2$)/yr]			Filtered plus particulate THg annual yields [($\mu g/m^2$)/yr]		
	Yield	Lower CI	Upper CI	Yield	Lower CI	Upper CI	Yield	Lower CI	Upper CI
McTier Creek, S.C.	1.357	1.056	1.710	0.729	0.465	1.082	2.086	1.522	2.792
Edisto River, S.C.	1.108	0.758	1.564	0.405	0.188	0.767	1.513	0.946	2.331
Fishing Brook, N.Y.	1.281	1.180	1.394	0.237	0.205	0.271	1.518	1.385	1.665
Hudson River, N.Y.	1.207	1.132	1.287	0.314	0.243	0.400	1.521	1.375	1.687
	Water year 2006								
McTier Creek, S.C.	1.019	0.843	1.220	0.472	0.351	0.620	1.490	1.194	1.840
Edisto River, S.C.	0.567	0.453	0.700	0.259	0.148	0.421	0.825	0.601	1.121
Fishing Brook, N.Y.	1.946	1.762	2.130	0.357	0.305	0.414	2.303	2.067	2.543
Hudson River, N.Y.	1.853	1.741	1.970	0.491	0.392	0.607	2.344	2.133	2.577
	Water year 2007								
McTier Creek, S.C.	0.994	0.755	1.270	0.381	0.269	0.526	1.375	1.024	1.796
Edisto River, S.C.	0.728	0.544	0.954	0.239	0.140	0.383	0.967	0.684	1.337
Fishing Brook, N.Y.	1.639	1.486	1.793	0.300	0.257	0.349	1.940	1.744	2.142
Hudson River, N.Y.	1.615	1.500	1.735	0.448	0.336	0.585	2.063	1.836	2.321
	Water year 2008								
McTier Creek, S.C.	0.663	0.548	0.792	0.252	0.189	0.314	0.914	0.737	1.107
Edisto River, S.C.	0.428	0.334	0.541	0.195	0.106	0.331	0.623	0.440	0.872
Fishing Brook, N.Y.	1.780	1.624	1.946	0.328	0.282	0.378	2.108	1.906	2.324
Hudson River, N.Y.	1.854	1.727	1.989	0.513	0.386	0.668	2.367	2.113	2.657
	Water year 2009								
McTier Creek, S.C.	1.005	0.792	1.253	0.434	0.309	0.591	1.439	1.102	1.844
Edisto River, S.C.	0.939	0.696	1.239	0.344	0.182	0.594	1.283	0.878	1.834
Fishing Brook, N.Y.	1.563	1.431	1.685	0.280	0.245	0.322	1.843	1.676	2.007
Hudson River, N.Y.	1.473	1.388	1.563	0.376	0.302	0.463	1.849	1.689	2.025

	5-Year mean annual yields [($\mu g/m^2$)/yr]			2-Year mean annual yields [($\mu g/m^2$)/yr]			5-Year lower CI			5-Year upper CI		
	Filtered THg	Particulate THg	THg	Filtered THg	Particulate THg	THg	Filtered THg	Particulate THg	THg	Filtered THg	Particulate THg	THg
McTier Creek, S.C.	1.007	0.454	1.461	0.834	0.343	1.177	0.799	0.317	1.116	1.249	0.627	1.876
Edisto River, S.C.	0.754	0.288	1.042	0.683	0.270	0.953	0.557	0.153	0.710	1.000	0.499	1.499
Fishing Brook, N.Y	1.642	0.301	1.942	1.672	0.304	1.976	1.497	0.259	1.756	1.790	0.347	2.136
Hudson River, N.Y.	1.601	0.428	2.029	1.664	0.444	2.108	1.498	0.332	1.829	1.709	0.545	2.253

Appendix 3-A. Calculated loads of fluvial total mercury (THg) and methylmercury (MeHg) for water years 2005 to 2009, with lower and upper bounds of the 95 percent confidence interval (CI) for the model.—Continued

[kg/yr, kilograms per year; $(\mu g/m^2)$/yr, micrograms per square meter per year. Total mercury includes methylated and inorganic forms of mercury. A water year extends from October of one calendar year to September of the following calendar year. For the McTier Creek and Fishing Brook sites, discharge records were extended to water years 2005, 2006, and 2007 using MOVE 1 estimation techniques; therefore, the 2005, 2006, and 2007 annual loads at these two sites were estimated using the extended discharge record Gaged discharge data were used for load computations for all annual loads at Edisto River and Hudson River sites and for 2008 and 2009 annual loads for the McTier Creek and Fishing Brook sites]

D. Methylmercury (MeHg) yields, by water year								
Site	**Water year 2005**							
	Filtered MeHg annual yields [($\mu g/m^2$)/yr]			**Particulate MeHg annual yields [($\mu g/m^2$)/yr]**			**Filtered plus particulate MeHg annual yields [($\mu g/m^2$)/yr]**		
	Yield	**Lower CI**	**Upper CI**	**Yield**	**Lower CI**	**Upper CI**	**Yield**	**Lower CI**	**Upper CI**
McTier Creek, S.C.	0.042	0.035	0.047	0.023	0.009	0.048	0.064	0.044	0.094
Edisto River, S.C.	0.129	0.077	0.202	0.019	0.011	0.033	0.148	0.087	0.235
Fishing Brook, N.Y.	0.074	0.061	0.089	0.012	0.011	0.015	0.086	0.072	0.104
Hudson River, N.Y.	0.050	0.044	0.058	0.012	0.009	0.016	0.063	0.053	0.074
Water year 2006									
McTier Creek, S.C.	0.030	0.028	0.034	0.013	0.006	0.020	0.043	0.034	0.054
Edisto River, S.C.	0.053	0.040	0.069	0.010	0.007	0.015	0.063	0.047	0.083
Fishing Brook, N.Y.	0.113	0.091	0.139	0.020	0.017	0.023	0.133	0.108	0.162
Hudson River, N.Y.	0.076	0.065	0.088	0.021	0.015	0.027	0.097	0.080	0.116
Water year 2007									
McTier Creek, S.C.	0.026	0.024	0.025	0.009	0.005	0.014	0.035	0.029	0.039
Edisto River, S.C.	0.058	0.042	0.077	0.009	0.006	0.013	0.067	0.048	0.090
Fishing Brook, N.Y.	0.081	0.066	0.098	0.014	0.012	0.017	0.095	0.079	0.115
Hudson River, N.Y.	0.054	0.046	0.064	0.016	0.011	0.022	0.070	0.057	0.086
Water year 2008									
McTier Creek, S.C.	0.021	0.020	0.024	0.006	0.004	0.009	0.028	0.024	0.033
Edisto River, S.C.	0.036	0.026	0.048	0.007	0.005	0.011	0.043	0.031	0.059
Fishing Brook, N.Y.	0.096	0.079	0.117	0.017	0.014	0.020	0.113	0.093	0.137
Hudson River, N.Y.	0.071	0.061	0.083	0.021	0.015	0.029	0.092	0.076	0.112
Water year 2009									
McTier Creek, S.C.	0.029	0.025	0.033	0.011	0.008	0.015	0.040	0.033	0.048
Edisto River, S.C.	0.091	0.063	0.128	0.015	0.009	0.023	0.106	0.072	0.150
Fishing Brook, N.Y.	0.093	0.077	0.112	0.015	0.013	0.018	0.108	0.089	0.130
Hudson River, N.Y.	0.064	0.055	0.074	0.016	0.012	0.020	0.080	0.067	0.094

	5-Year mean annual yields [($\mu g/m^2$)/yr]			**2-Year mean annual yields [($\mu g/m^2$)/yr]**			**5-Year lower CI**			**5-Year upper CI**		
	Filtered MeHg	**Particulate MeHg**	**MeHg**	**Filtered MeHg**	**Particulate MeHg**	**MeHg**	**Filtered MeHg**	**Particulate MeHg**	**MeHg**	**Filtered MeHg**	**Particulate MeHg**	**MeHg**
McTier Creek, S.C.	0.030	0.012	0.042	0.025	0.009	0.034	0.026	0.006	0.033	0.032	0.021	0.054
Edisto River, S.C.	0.073	0.012	0.085	0.064	0.011	0.075	0.050	0.008	0.057	0.105	0.019	0.123
Fishing Brook, N.Y.	0.092	0.016	0.107	0.095	0.016	0.111	0.075	0.013	0.088	0.111	0.019	0.130
Hudson River, N.Y.	0.063	0.017	0.080	0.068	0.018	0.086	0.054	0.013	0.067	0.073	0.023	0.096

Appendix 3-B. S–LOADEST model coefficients and estimated annual loads of mercury, dissolved organic carbon, suspended sediment, and major ions for Fishing Brook at County Line Flow near Newcomb, N.Y. (station 0131199050), McTier Creek near New Holland, S.C. (station 02172305), McTier Creek near New Holland, S.C. (station 02172305), Edisto River near Givhans, S.C. (station 02175000), and Hudson River near Newcomb, N.Y. (station 01312000) for water years 2005 to 2009.

[For the McTier Creek and Fishing Brook sites, discharge records were extended to water years 2005, 2006, and 2007 using MOVE.1 estimation techniques; therefore, the 2005, 2006, and 2007 annual loads at these two sites were estimated using the extended discharge record. Gaged discharge data were used for load computations for all annual loads at the Edisto River and Hudson River sites and for 2008 and 2009 annual loads for the McTier Creek and Fishing Brook sites. FTHg, filtered total mercury; FMeHg, filtered methylmercury; PTHg, particulate total mercury; PMeHg, particulate methylmercury; DOC, dissolved organic carbon; POC, particulate organic carbon; SSC, suspended sediment; mg/yr, milligrams per year; (µg/m²)/yr, micrograms per square meter per year; (g/m²)/yr, grams per square meter per year; kg/yr, kilograms per year; g/yr, grams per square meter per year. A water year extends from October of one calendar year to September of the following calendar year]

Parameter	n (cen)	R-squared	β_0 Intercept	β_1 LnQ	β_2 LnQ²	β_3 sin DecTime	β_4 cos DecTime	Annual load (mg/yr) WY 2005	WY 2006	WY 2007	WY 2008	WY 2009	Mean annual load (mg/yr)	Mean annual yield [(µg/m²)/yr]	Annual load (mg/yr) WY 2008	WY 2009	Annual yield [(µg/m²)/yr] WY 2008	WY 2009
								McTier Creek near New Holland, S.C. (02172305)										
FMeHg	45 (4)	87.26	1.630	0.842	-0.005	0.293	-0.536	3,262	2,430	2,132	1,749	2,295	2,374	0.030	1,749	2,295	0.022	0.029
FTHg	43 (0)	88.64	4.811	1.193	0.109	0.229	0.166	107,874	80,735	78,811	52,717	79,901	80,008	1.006	52,717	79,901	0.663	1.005
PTHg	45 (1)	78.86	4.005	1.412	0.057	-0.194	-0.393	57,958	37,513	30,343	9,987	34,505	36,061	0.454	19,987	34,505	0.251	0.434
PMeHg	45 (13)	69.64	0.105	1.343	0.151	-0.044	-1.037	1,832	962.1	701.3	488.6	850.4	966.9	0.012	489	850	0.006	0.011
								Edisto River near Givhans, S.C. (02175000)										
FMeHg	25 (0)	87.62	6.622	1.784	-0.220	-0.328	-0.520	908,658	375,119	407,072	254,979	643,545	517,875	0.073	254,979	643,545	0.036	0.091
FTHg	24 (0)	91.79	8.794	1.523	0.113	-0.065	-0.061	7,833,704	4,007,060	5,145,718	3,025,660	6,636,684	5,329,765	0.754	3,025,660	6,636,684	0.428	0.939
PTHg	25 (1)	65.99	8.225	1.233	-0.528	0.200	-0.623	2,861,336	1,827,679	1,693,399	1,381,266	2,433,963	2,039,529	0.288	1,381,266	2,433,963	0.195	0.344
PMeHg	25 (1)	81.79	5.095	1.479	-0.667	0.075	-0.939	137,403	71,379	63,776	50,387	105,974	85,784	0.012	50,387	105,974	0.007	0.015
								Fishing Brook (County Line Flow) near Newcomb, N.Y. (0131119950)										
FMeHg	41 (6)	79.52	2.730	0.846	-0.169	-0.414	-0.465	4,834	7,373	5,299	6,291	6,083	5,976	0.092	6,291	6,082	0.096	0.093
FTHg	41 (0)	97.62	5.347	1.163	-0.011	-0.112	-0.293	83,593	126,849	106,859	116,198	101,671	107,034	1.640	116,198	101,671	1.780	1.558
PTHg	41 (0)	93.12	3.601	1.086	0.056	-0.218	-0.355	15,460	23,279	19,618	21,363	18,280	19,600	0.300	21,363	18,280	0.327	0.280
PMeHg	41 (8)	88.2	0.762	0.954	0.001	-0.482	-0.706	833.7	1,294	937.2	1,098	984	1,030	0.016	1,098	984	0.017	0.015
								Hudson River near Newcomb, N.Y. (01312000)										
FMeHg	32 (4)	84.58	4.458	0.774	-0.141	-0.453	-0.482	25,020	37,778	27,084	35,480	31,811	31,435	0.063	35,480	31,811	0.071	0.064
FTHg	32 (0)	98.81	7.418	1.142	-0.018	-0.109	-0.070	600,430	921,444	803,004	922,442	732,708	796,006	1.601	922,442	732,709	1.855	1.473
PTHg	32 (0)	92.44	5.831	1.393	-0.017	-0.112	-0.052	156,082	244,161	222,924	254,820	186,910	212,979	0.428	254,817	186,910	0.512	0.376
PMeHg	32 (9)	87.08	2.676	1.351	-0.074	-0.369	-0.523	6,088	10,330	7,918	10,390	7,804	8,506	0.017	10,390	7,804	0.021	0.016

Appendix 3-B. S-LOADEST model coefficients and estimated annual loads of mercury, dissolved organic carbon, suspended sediment, and major ions for Fishing Brook at County Line Flow near Newcomb, N.Y. (station 0131199050), McTier Creek near New Holland, S.C. (station 02172305), McTier Creek near New Holland, S.C. (station 02172305), Edisto River near Givhans, S.C. (station 02175000), and Hudson River near Newcomb, N.Y. (station 01312000) for water years 2005 to 2009.—Continued

[For the McTier Creek and Fishing Brook sites, discharge records were extended to water years 2005, 2006, and 2007 using MOVE.1 estimation techniques; therefore, the 2005, 2006, and 2007 annual loads at these two sites were estimated using the extended discharge record. Gaged discharge data were used for load computations for all annual loads at the Edisto River and Hudson River sites and for 2008 and 2009 annual loads for the McTier Creek and Fishing Brook sites. FTHg, filtered total mercury; FMeHg, filtered methylmercury; PTHg, particulate total mercury; PMeHg, particulate methylmercury; DOC, dissolved organic carbon; POC, particulate organic carbon; SSC, suspended sediment; mg/yr, milligrams per year; (µg/m²)/yr, micrograms per square meter per year; (g/m²)/yr, grams per square meter per year; kg/yr, kilograms per year; (g/m²)/yr, grams per square meter per year. A water year extends from October of one calendar year to September of the following calendar year]

Parameter	n (cen)	R-squared	β_0 Intercept	β_1 LnQ	β_2 LnQ²	β_3 sin. DecTime	β_4 cos. DecTime	Annual load (kg/yr) WY 2005	WY 2006	WY 2007	WY 2008	WY 2009	Mean annual load (kg/yr)	Mean annual yield [(g/m²)/yr]	Annual load (kg/yr) WY 2008	WY 2009	Annual yield [(g/m²)/yr] WY 2008	WY 2009
McTier Creek near New Holland, S.C. (02172305)																		
DOC	43 (0)	98.84	5.810	1.292	0.104	-0.062	-0.188	293,472	197,988	176,019	113,831	188,904	194,043	2.440	113,831	188,904	1.432	2.376
SSC	44 (0)	85.57	5.850	1.722	-0.081	-0.566	-0.888	472,135	261,609	165,708	107,639	225,278	246,474	3.100	107,639	225,278	1.354	2.833
POC	40 (0)	84.28	4.209	1.386	-0.083	-0.553	-0.742	56,581	36,038	24,656	17,535	29,852	32,932	0.414	17,535	29,852	0.221	0.375
Chloride	45 (0)	99.55	5.133	0.950	-0.009	-0.021	0.088	98,896	81,474	72,061	55,518	74,000	76,390	0.961	55,518	74,000	0.698	0.931
Sulfate	45 (0)	97.16	4.152	1.610	-0.084	0.054	0.160	58,059	43,755	40,022	25,576	41,541	41,791	0.526	25,576	41,541	0.322	0.522
Edisto River near Givhans, S.C. (02175000)																		
DOC	44 (0)	96.25	10.898	1.341	-0.077	-0.097	-0.066	16,820,843	9,575,338	11,165,734	7,038,295	14,357,131	11,791,468	1.668	7,038,295	14,357,131	0.995	2.031
SSC	58 (0)	68.78	10.003	1.208	-0.535	0.086	-0.460	14,584,853	9,436,573	8,742,129	6,891,934	12,154,648	9,913,872	1.402	6,891,934	12,154,648	0.975	1.719
POC	11 (0)	ND	ND	ND	ND	ND	ND	ND	ND	ND	ND	ND	ND	ND	ND	ND	ND	ND
Chloride	123 (0)	94.4	10.461	0.749	-0.001	0.064	0.084	12,514,659	9,547,558	10,090,493	7,780,359	11,606,374	10,307,889	1.458	7,780,359	11,606,374	1.100	1.641
Sulfate	123 (0)	72.39	10.080	0.667	0.014	-0.128	0.123	8,892,109	7,059,935	7,307,273	5,687,131	822,700	5,953,830	0.842	5,687,131	822,700	0.804	0.116
Fishing Brook (County Line Flow) near Newcomb, N.Y. (013119950)																		
DOC	39 (0)	98.75	6.703	1.065	-0.006	-0.300	-0.229	274,444	429,459	353,540	378,921	339,321	355,137	5.441	378,921	339,321	5.806	5.199
SSC	39 (2)	87.24	3.553	0.906	0.025	-0.567	-0.751	13,310	20,481	14,712	17,395	15,498	16,279	0.249	17,395	15,498	0.267	0.237
POC	39 (4)	87.28	3.535	0.908	0.029	-0.561	-0.759	13,181	20,260	14,569	17,219	15,313	16,108	0.247	17,219	15,313	0.264	0.235
Chloride	39 (0)	93.35	6.430	0.897	-0.032	-0.123	0.069	201,908	297,281	254,168	270,585	253,571	255,503	3.915	270,585	253,571	4.146	3.885
Sulfate	39 (00)	99.3	6.063	0.965	-0.025	0.110	0.235	149,180	214,560	192,451	205,497	188,643	190,066	2.912	205,497	188,643	3.149	2.890
Hudson River near Newcomb, N.Y. (01312000)																		
DOC	35 (0)	97.95	8.324	1.114	-0.030	-0.296	-0.067	1,957,978	3,137,351	2,645,087	2,997,565	2,412,050	2,630,006	5.289	2,997,565	2,412,050	6.028	4.851
SSC	36 (5)	77.59	7.069	1.239	0.262	-0.244	-0.034	780,891	1,123,795	1,174,291	1,321,893	810,055	1,042,185	2.096	1,321,893	810,055	2.658	1.629
POC	32 (5)	93.24	5.266	1.406	-0.039	-0.430	-0.465	81,584	140,496	111,571	143,020	101,665	115,667	0.233	143,020	101,665	0.288	0.204
Chloride	32 (0)	96.18	7.881	0.770	0.082	0.144	0.075	992,674	1,313,374	1,212,280	1,353,842	1,140,103	1,202,455	2.418	1,353,842	1,140,103	2.723	2.293
Sulfate	32 (0)	96.88	8.712	0.740	0.017	0.005	0.018	2,004,752	2,750,271	2,397,285	2,692,071	2,362,345	2,441,345	4.909	2,692,071	2,362,345	5.414	4.751

Appendix 3-E. Estimated annual wet deposition of major ions and nutrients in the Edisto River basin in South Carolina based on monitoring at the National Atmospheric Deposition Program National Trend Network (NADP NTN) site SC06 and in the upper Hudson River basin based on monitoring at the NADP NTN site NY20 for water years 2005 to 2009 (see table 1 for site details or access *http://nadp.sws.uiuc.edu/data/ntndata.aspx*)

[A water year extends from October 1 of one calendar year to September 30 of the next calendar year; (kg/ha)/yr, kilograms per hectare per year; ml, milliliter; cm, centimeter; %, percent; for more information on data completeness criteria, please see the NADP Web site]

| Site ID | Water year | Data completeness criteria (%) | | | | Wet deposition [(kg/ha)/yr] | | | | | | | | | | Totals | | | | Dates |
		1	2	3	4	Calcium	Magnesium	Potassium	Sodium	Ammonia	Nitrate	Inorganic nitrogen	Chloride	Sulfate	Hydrogen ion	Sample volume (ml)	Precipitation (cm)	Valid samples	Days	
SC06	2005	83	96	98	95	0.58	0.182	0.228	1.341	1.44	6.88	2.67	2.43	11.69	0.24	59803.6	91.23	36	364	9/28/2004 9/27/2005
SC06	2006	77	96	95	97	1.16	0.369	0.449	2.386	1.86	8.51	3.36	4.37	14.65	0.28	72513.1	115.26	34	371	9/27/2005 10/3/2006
SC06	2007	90	100	93	100	0.79	0.31	0.382	2.254	1.53	7.26	2.83	4.14	12.26	0.24	64668.4	103.38	33	364	10/3/2006 10/2/2007
SC06	2008	88	100	95	101	1.03	0.33	0.408	1.923	2.03	7.62	3.3	3.54	14.94	0.25	63501.8	97.11	33	364	10/2/2007 9/30/2008
SC06	2009	94	100	97	101	1	0.346	0.466	2.304	1.79	6.23	2.8	4.01	9.23	0.14	80670.3	119.4	36	364	9/30/2008 9/29/2009
	Mean	86.4	98.4	95.6	99.6	0.912	0.307	0.387	2.042	1.73	7.3	2.992	3.708	12.554	0.23	68231.44	105.276	34.4	365.4	
NY20	2005	98	100	100	90	0.55	0.087	0.141	0.271	1.62	10.31	3.59	0.72	11.33	0.28	66036.6	108.43	49	364	9/28/2004 9/27/2005
NY20	2006	94	100	96	92	0.83	0.161	0.269	0.632	1.9	9.74	3.67	1.2	12.21	0.26	80279.5	134.47	50	371	9/27/2005 10/3/2006
NY20	2007	88	100	82	91	0.84	0.129	0.097	0.259	2.21	9.43	3.85	0.6	12.47	0.25	55408.9	107.78	44	364	10/3/2006 10/2/2007
NY20	2008	87	100	86	91	0.91	0.126	0.126	0.294	2.46	11.7	4.56	0.66	15.08	0.33	74626	139.78	43	364	10/2/2007 9/30/2008
NY20	2009	88	100	87	96	0.71	0.09	0.113	0.248	1.78	7.96	3.18	0.56	8.23	0.18	64058	112.74	45	364	9/30/2008 9/29/2009
	Mean	91	100	90.2	92	0.768	0.119	0.149	0.341	1.994	9.828	3.77	0.75	11.864	0.26	68081.8	120.64	46.2	365.4